全　智　编著

现代多用户和多天线检测技术

Modern Multiuser and Multiple-Antenna Detection Techniques

郑州大学出版社
郑州

图书在版编目(CIP)数据

现代多用户和多天线检测技术＝Modern Multiuser and Multiple-Antenna Detection Techniques:汉英对照/全智编著.
—郑州:郑州大学出版社,2018.10
ISBN 978-7-5645-5828-4

Ⅰ.①现…　Ⅱ.①全…　Ⅲ.①码分多址-汉、英
Ⅳ.①TN914.53

中国版本图书馆 CIP 数据核字(2018)第 214602 号

郑州大学出版社出版发行
郑州市大学路 40 号　　　　邮政编码:450052
出版人:张功员　　　　发行部电话:0371-66966070
全国新华书店经销
河南瑞之光印刷股份有限公司印制
开本:787 mm×1 092 mm　1/16
印张:13
字数:371 千字
版次:2018 年 10 月第 1 版　　　　印次:2018 年 10 月第 1 次印刷

书号:ISBN 978-7-5645-5828-4　　　　定价:198.00 元

Preface

Detection for multiuser and multiple-input multiple-out (MIMO) systems is a key technology that can enhance a communication system's spectral efficiency and anti-interference performance. In the communication networks of the future, when base stations will deploy 100 or more antennas to simultaneously serve multiple users, achieving reliable signal detection will be one of the main problems that must be resolved. However, while maximum likelihood detector (MLD) can provide optimal performance because the amount of needed computing power increases exponentially with increasing system size, it is difficult to use MLD in practical operations. In addition, the use of linear detection algorithms, such as the zero forcing and minimum mean square error detection algorithms, can reduce complexity. However their performance is considerably inferior to that of an optimal detection algorithm. Because linear detection algorithms must calculate matrix inversion, the computing load remains large as the system size increases. In contrast, because dichotomous coordinate descent (DCD) detection algorithm does not require multiplications/divisions operations, it is highly suitable for hardware implementation. In next-generation 5G wireless communications, even more Internet of Things equipment will be incorporated in cellular networks. As a consequence, the design and development of superior high-throughput, low-complexity detection algorithms are urgently needed. Accordingly, the author, based on many years of overseas study and work experience, and after referring to many foreign books, has written *Modern Multiuser and Multiple-Antenna Detection Techniques* as a practical reference for people engaging in electronic technology and communications. In addition, numerous universities in China have offered professional courses taught in English; however, these courses are typically a poor fit with professional subject matter and use teaching materials with a format inconsistent with foreign styles. Thus, this book can serve as a bilingual or all-English textbook for postgraduate students in the fields of electronics, information, and communications.

This book provides a detailed introduction to the design of multi-user and multi-antenna detection algorithms, as well as relevant hardware implementation, from the perspective of professional practice. The book consists of four sections. The first section (chapters 1 ~ 3) introduces code division multiple access (CDMA) technology and the establishment of channel models. The second section (Chapter 4) concerns the use of linear matrix equations to solve problems involving signal processing applications and provides an overview of hardware implementation. The third section (chapters 5 ~ 6) introduces multiuser detection and MIMO technologies. The fourth section (chapters 7 ~ 8) chiefly introduces DCD based multiuser and MIMO detection algorithms design and hardware implementations. In this book, the content of the first and second sections provides a review of the professional knowledge learned at the undergraduate stage and has the goal of providing relevant specialized English vocabulary to readers learning professional knowledge in the field so that they may accurately understand the book's content. At the same time, readers can rely on their existing professional knowledge to verify the theory and mathematical deductions presented in the book. The content of the third section is about the theoretical knowledge and application technology in the area of multiuser and MIMO systems that is seldom taught in university classrooms but it is needed by professionals in relevant fields. The fourth section presents the practical applications of DCD algorithms in multiuser and MIMO detection.

In vicw of the gaps that may exist in readers' basic knowledge, particularly in English reading ability, apart from proceeding from the easier aspects to the more complicated material and striving to present concepts in the most concise and accurate manner, Chinese notes corresponding to the content of each chapter have been included at the end of the book. This will help readers better understand key points in the text, while also familiarizing them with professional international English expression and writing habits in this field. As a result, readers will be able to quickly understand and follow the cutting edge of development in this field, and the book will help train technical personnel possessing international competitiveness.

During the process of writting this work, Dr. Yuriy provided numerous excellent suggestions and revised the book's final English manuscript. I am also grateful to Chaoyi Ma, Shuhua Lv, and Shanshan Li, who drew some figures for the book.

Although the author put great effort into writing this book, because of possible difficult-to-avoid errors and shortcomings attributable to the author's limited knowledge and ability, I invite readers to provide their comments, which will help me make revisions in further editions.

前 言

多用户和多输入多输出(multiple input multiple output, MIMO)检测是增强通信系统抗干扰性和提高频谱效率的关键技术。在未来通信网络中,基站配置百根天线同时服务多个用户,可靠的信号检测将是面临的主要问题之一。最大似然检测算法可以提供最优性能,但是运算量随系统规模呈指数增长,因此难以应用于实际工作。线性检测算法,例如迫零、最小均方误差等次优检测算法,虽然降低了复杂度,但是性能与最优检测算法仍有差距。另外,随着系统规模的增大,这些线性检测算法由于需要计算矩阵求逆,运算负担仍然很重。基于二分坐标下降(dichotomous coordinate descent, DCD)的检测算法不需要乘法/除法操作,非常适合硬件实施。在下一代的5G无线通信中,将会有更多物联网设备接入蜂窝网络。因此迫切需要设计和开发高吞吐量和低复杂度的优良检测算法。为此,编者根据多年的国外留学和工作经验,参考了大量国外原版图书,编写了这部《现代多用户和多天线检测技术》,供广大从事电子技术和通信专业的同行在实际工作中参考。另外,基于我国不少高等院校都开设了专业英语课程,但普遍缺乏专业契合度高且文体形式与国外接轨教材的现状,该书也可作为电子信息与通信类专业研究生双语和全英语教学的教材。

本书从专业人员的工作实际出发,较为详细地介绍了多用户和多天线检测算法的设计和硬件实施。全书可分为4个部分。第一部分(第1~3章)介绍码分多址(code division multiple access, CDMA)技术和信道模型的建立;第二部分(第4章)是运用线性矩阵方程解决实际信息处理应用中的问题以及硬件实施概述;第三部分(第5~6章)介绍多用户检测和MIMO系统技术;

第四部分(第7～8章)主要介绍DCD算法的多用户和MIMO检测算法的设计和硬件实现。其中,第一部分和第二部分的内容是对大学本科阶段相关专业知识的回顾,目的是使读者在具有专业知识的背景中学习、掌握相关英语专业词汇,以便准确理解文中内容。同时,利用已有的专业知识来验证相关章节中的理论。第三部分内容是大学课堂教学很少涉足的,但也是相关领域从业人员需要学习和掌握的关于多用户及MIMO系统的理论知识和应用技术。第四部分是DCD类算法在多用户和MIMO检测中的实际应用。

考虑到读者的基础水平尤其是英语阅读能力存在差异,本书在编写过程中除注意到在内容上由浅入深、概念力求简洁准确外,还在书后附有与每章内容对应的中文注释,以帮助读者更方便地理解文中的一些关键知识点的内涵,同时熟悉该领域专业英语的国际通用表达和书写习惯,使读者能及时掌握和跟踪该领域发展前沿,为行业培养具有国际竞争能力的技术人才。

在编写该书的过程中,Dr. Yuriy曾给了不少好的建议,并对本书的最终英文文稿做了修改;马超艺、吕树花、李姗珊绘制了部分图表。在此一并表示感谢。

尽管编者为本书倾注了很大的精力,但由于个人的水平和能力有限,书中的错误、缺点在所难免,欢迎广大读者不吝赐教,把您的意见和建议反馈给我,以利于再版时修正。

全　智

2018年4月

目 录

PART Ⅰ

PART Ⅱ

PART Ⅲ

PART Ⅳ

第一部分

第二部分

第三部分

第四部分

Chapter 1 Spread-spectrum technology

1.1 Signature waveforms

1.1.1 Spread spectrum

Spread-spectrum (SS) communications technology was first described on paper by an actress and a musician. In 1941, Hollywood actress Hedy Lamarr and pianist George Antheil described a secure radio link to control torpedoes and received U. S. patent #2. 292. 387. The technology was not taken seriously by the U. S. Army at that time and was forgotten until the 1980s, when resurfaced and has become increasingly popular for applications that involve radio links in hostile environments[1].

Typical applications for the resulting short-range data transceivers include satellite-positioning systems (GPS), 3G mobile telecommunications, W-LAN (IEEE802. 11a, IEEE802. 11b, IEE802. 11g), and Bluetooth. SS techniques also aid in the endless race between communication needs and radio-frequency availability (the radio spectrum is limited and is therefore an expensive resource).

SS is apparent in the Shannon and Hartley channel-capacity theorem:

$$C = B \cdot \log_2(1 + S/N) \tag{1.1}$$

In this equation, C is the channel capacity in bits per second (bps), which is the maximum data rate for a theoretical bit error rate (BER). B is the required channel bandwidth in Hz, and S/N is the signal-to-noise power ratio. Specifically, one assumes that C, which represents the amount of information allowed by the communication channel, also represents the desired performance. Bandwidth is the price paid because frequency is a limited resource. The S/N ratio expresses the environmental conditions or the physical characteristics (obstacles, presence of jammers, interference, etc.).

An elegant interpretation of this equation, applicable for difficult environments (low S/N ratios due to noise and interference), says that one can maintain or even increase communication performance (high C) by allowing or injecting more bandwidth (high B), even when signal power is below the noise floor.

The above equation can be modified by changing the log base from 2 to e (the Napierian logarithm), and by noting that $\ln = \log_e$

$$C/B = (1/\ln 2) \cdot \ln(1 + S/N) = 1.443 \cdot \ln(1 + S/N) \tag{1.2}$$

Applying the MacLaurin series development for

$$\ln(1 + x) = x - x^2/2 + x^3/3 - x^4/4 + \cdots + (-1)^{k+1}x^k/k + \cdots \tag{1.3}$$

$$C/B = 1.443 \cdot [S/N - 1/2 \cdot (S/N)^2 + 1/3 \cdot (S/N)^3 - \cdots] \tag{1.4}$$

S/N is usually low for spread-spectrum applications. (As just mentioned, the signal power density can even be below the noise level.) Assuming a noise level such that $S/N \ll 1$, Shannon´s expression becomes: $C/B \approx 1.443 \cdot S/N$. Very roughly,

$$C/B \approx S/N \quad \text{or} \quad N/S \approx B/C \tag{1.5}$$

Therefore, we need to perform only the fundamental SS signal-spreading operation to send error-free information for a given noise-to-signal ratio in the channel: increase the transmitted bandwidth. This principle appears simple and evident, but its implementation is complex mainly because spreading the baseband (by a factor that can be several orders of magnitude) forces the electronics to act and react accordingly, making spreading and de-spreading operations necessary.

Different SS techniques are available, but all have one idea in common: the key (also called code or sequence) attached to the communication channel. The manner of inserting this code precisely defines the SS technique in question. The term "spread spectrum" refers to the expansion of signal bandwidth, by several orders of magnitude in some cases, that occurs when a key is attached to the communication channel.

Spread spectrum is an radio frequency (RF) communications system (Figure 1 – 1) in which the baseband signal bandwidth is intentionally spread over a larger bandwidth by injecting a higher-frequency signal. As a direct consequence, energy used in transmitting the signal is spread over a wider bandwidth and appears as noise. The ratio (in dB) between the spread baseband and the original signal is called the processing gain. Typical SS processing gains range from 10 dB to 60 dB.

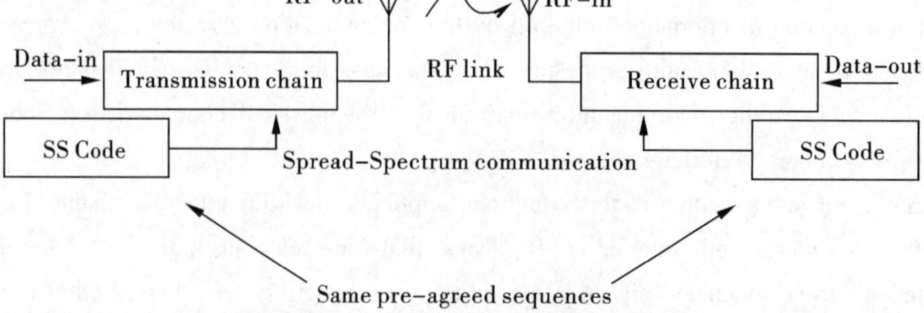

Figure 1-1 Spread-spectrum communication

Spreading factor

The choice of the number of chips persymbol N , also known as the spreading factor, spreading gain, or processing gain, has the following effects:

1) For fixed duration of the signature waveform, the bandwidth is proportionalto N .

2) For a given signal-to-noise ratio, the single-user BER in a white Gaussian noise channel is independentof N .

3) In an orthogonal synchronous direct-sequence SS (DSSS) code-division multiple access (CDMA) system, the number of users that can be supported is less than or equalto N .

4) Large values of N contribute to the privacy of the system as they hinder unintended receivers wishing to unveil signature waveform to eavesdrop on the transmitted information.

1.1.2 Bandwidth effects of the spreading operation

The simple drawings (Figure 1-2) below illustrate the evaluation of signal bandwidth in a communication link.

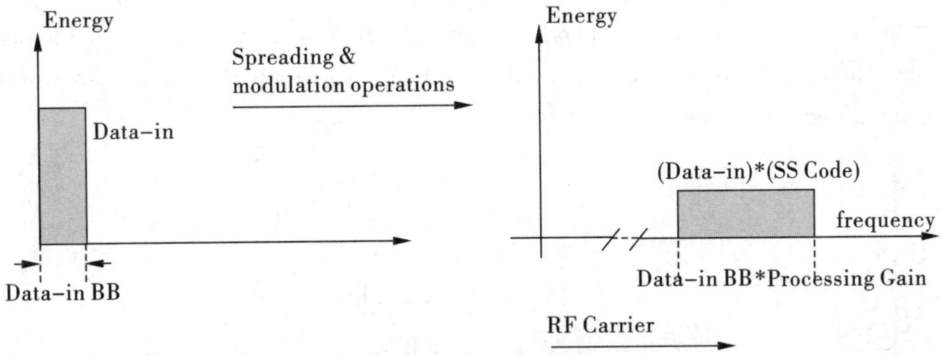

Figure 1-2 Spreading and modulation operations

SS modulation is applied on top of a conventional modulation such as binary phase shift keying (BPSK) or direct conversion. One can demonstrate that all other signals not receiving the SS code will remain un-spread.

1.1.3 Bandwidth effects of the de-spreading operation

Similarly, de-spreading is illustrated in the following figure (Figure 1-3).

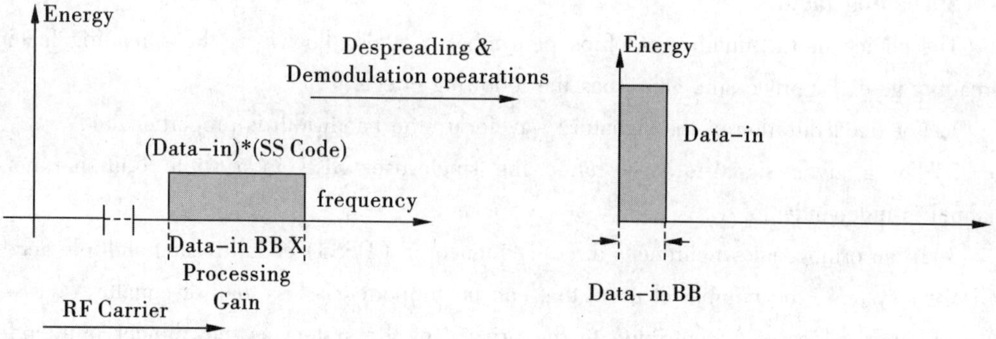

Figure 1-3 De-spreading and demodulation operations

An SS demodulation has been made on top of the normal demodulation operations above. One can also demonstrate that signals added during the transmission (such as an interference or jammer) will be spread during the de-spreading operation.

1.1.4 Waste of bandwidth due to spreading is off-set by multiple users

Spreading directly results in the use of a wider frequency band (by a factor corresponding exactly to the "processing gain" mentioned earlier), so it does not spare the limited frequency resource. However this overuse is compensated for by the possibility that many users will share the enlarged frequency band (Figure 1-4).

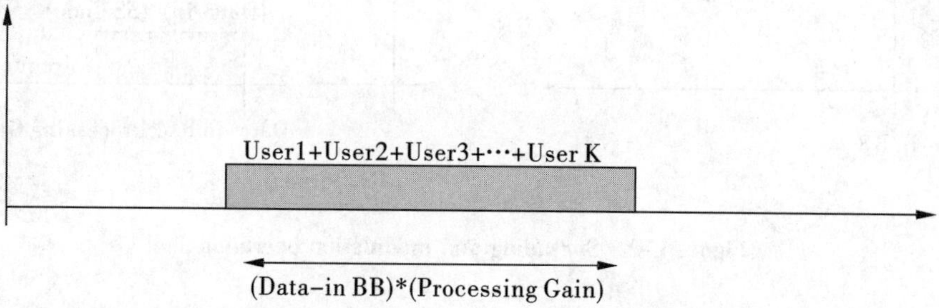

Figure 1-4 Enlarged frequency band shared by multiple users

As opposed to regular narrowband technology, the SS process of spreading is wideband technology. W-CDMA and the universal mobile telecommunications system (UMTS), for example, are wideband technologies that require a relatively large frequency bandwidth (compared to that of narrowband radio).

1.1.5 Resistance to interference and anti-jamming effects

Resistance to interference and anti-jamming are the major advantages SS. Intentional or un-intentional interference and jamming signals are rejected because they do not contain the SS key. Only the desired signal, which has the key, is seen at the receiver when de-spreading is performed (Figure 1-5).

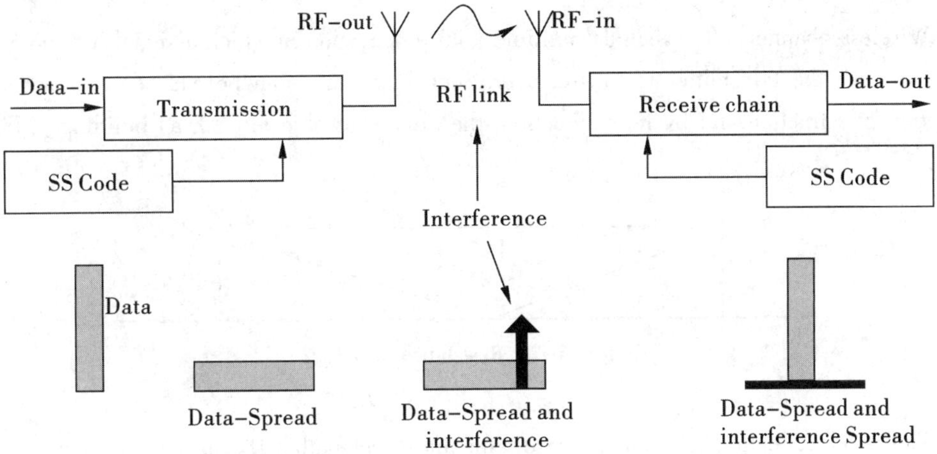

Figure 1-5 Anti-jamming by using SS key

The interference (narrowband or wideband) can essentially be ignored if it does not include the key used in the de-spreading operation. That rejection also applies to other SS signals that do not have the correct key, which allows different SS communications to be active simultaneously in the same band (such as CDMA). Note that SS is a wideband technology, but not all wideband techniques require SS technology.

1.1.6 Resistance to interception

Resistance to interception is the second advantage provided by SS techniques (Figure 1-6). Because non-authorized listeners do not have the key used to spread the original signal, they cannot decode it. Without the right key, the SS signal appears as noise or as an interferer (Scanning methods can break the code, however, if the key is short). Furthermore, the signal levels can be below the noise floor because the spreading operation reduces the spectral density (the total energy is the same, but it is widely spread in frequency). The message is thus made invisible, an effect that is particularly strong with the direct sequence spread spectrum technique. Other receivers cannot "see" the transmission; they only register a slight increase in the overall noise level.

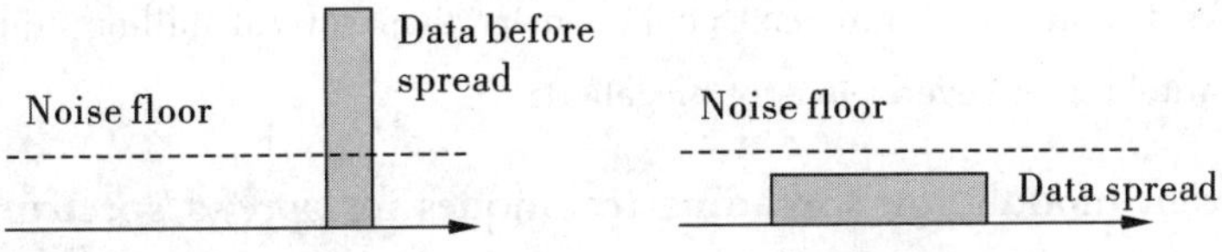

Figure 1-6 Resistance to interception

1.1.7 Resistance to fading (multipath effects)

Wireless channels often include multiple-path propagation in which a signal has more than one path from the transmitter to the receiver. Such multipaths can be caused by atmospheric reflection or refraction and by reflection from the ground or objects such as buildings (Figure 1-7).

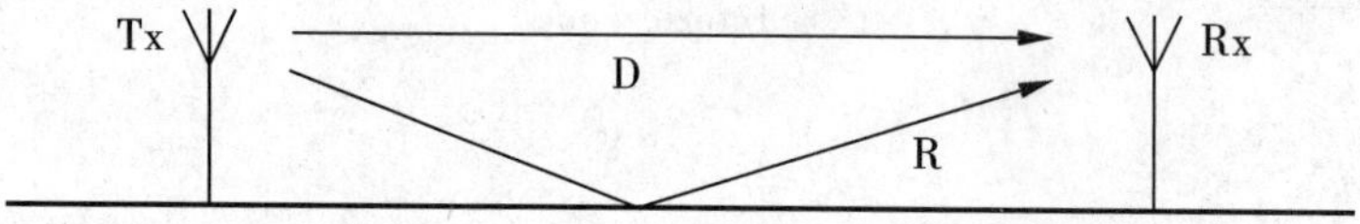

Figure 1-7 Resistance to fading

The reflected path (R) can interfere with the direct path (D) in a phenomenon called fading. Because the de-spreading process synchronizes to signal D, signal R is rejected even though it contains the same key. Methods are available to use the reflected-path signals by de-spreading them and adding the extracted results to the main result.

1.1.8 Spread spectrum and (de)coding "keys"

At this point, we know that the main SS characteristic is the presence of a code or key, which must be known in advance by the transmitter and receiver (s). In modern communications, the codes are digital sequences that must be as long and as random as possible to appear as "noise-like" as possible. However, they must also remain reproducible. Otherwise, the receiver will be unable to extract the message that has been sent. Thus, the sequence is " nearly random". Such a code is called a pseudo-random number (PRN) or sequence. The most frequently used method to generate pseudo-random codes is based on a feedback shift register.

To guarantee efficient SS communications, the PRN sequences must follow certain rules, such as length, autocorrelation, cross-correlation, orthogonality, and bits balancing. The more popular PRN sequences have names: Barker, M-Sequence, Gold, Hadamard-Walsh, etc. Note that a more complex sequence set provides a more robust SS link. However, the cost is more complex electronics (both in speed and behavior), mainly for the SS de-spreading operations. Purely digital SS de-spreading chips can contain more than several million equivalent 2-input NAND gates, switching at several tens of megahertz.

1.1.9 Different modulation spreading techniques for spread spectrum

Different SS techniques are distinguished according to the point in the system at which a PRN is inserted into the communication channel, as illustrated in the RF front end schematic shown in figure 1-8.

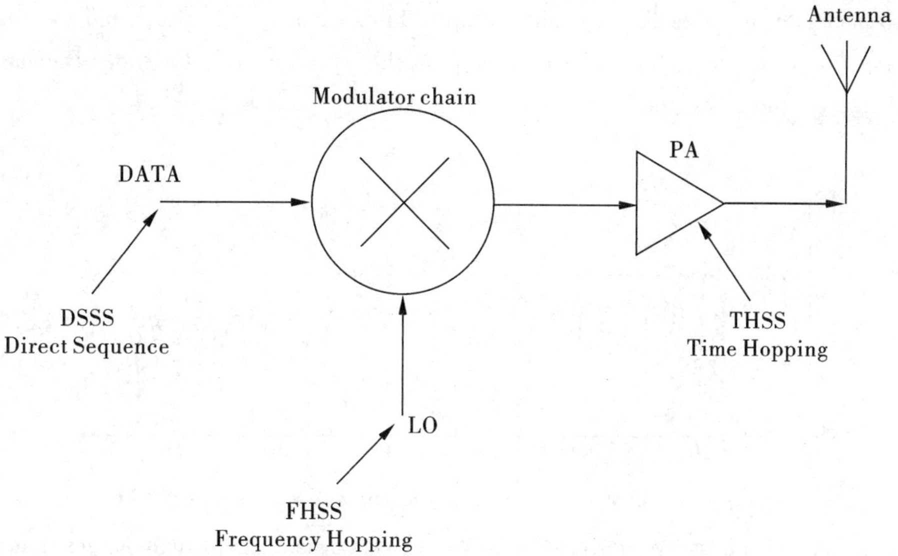

Figure 1-8 Different SS techniques in different points

If the PRN is inserted at the data level, we have DSSS. (In practice, the pseudo-random sequence is mixed or multiplied with the information signal, giving an impression that the original data flow was "hashed" by the PRN.) If the PRN acts at the carrier-frequency level, we have the frequency hopping form of spread spectrum (FHSS). Applied at the local oscillator (LO) stage, FHSS PRN codes force the carrier to change or hop according to the pseudo-random sequence. If the PRN acts as an on/off gate to the transmitted signal, we have a time hopping spread-spectrum technique (THSS). One can mix these techniques to form hybrid SS techniques, such as DSSS + FHSS. Currently, DSSS and FHSS are the two most commonly used techniques.

(1) **Direct-sequence spread spectrum (DSSS)**

In this technique, the PRN is applied directly to data entering the carrier modulator. The modulator therefore sees a much larger bit rate, which corresponds to the chip rate of the PRN sequence. The result of modulating an RF carrier with such a code sequence is to produce a direct-sequence-modulated spread spectrum with $[\sin(x)/x]^2$ frequency spectrum centered at the carrier frequency. The main lobe of this spectrum has a bandwidth twice that is the clock rate of the modulating code, and the side lobes have bandwidths equal to the code's clock rate[2].

Consider an example of this binary signal with a symbol rate of 2 bps (Figure 1-9).

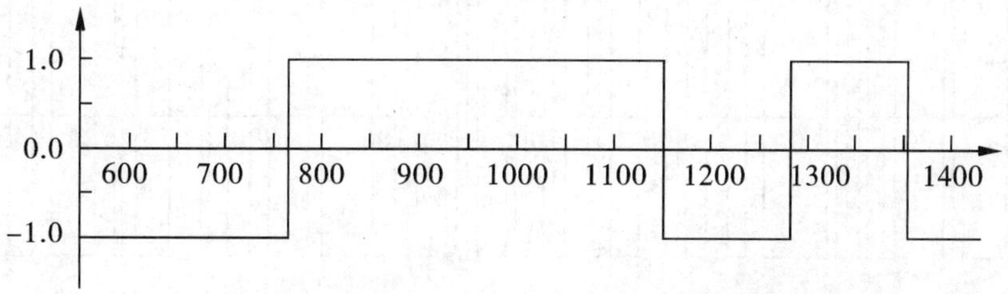

Figure 1-9 A binary information signal

To modulate this signal, we could multiply this sequence with a sinusoid to obtain the spectrum shown in figure 1–10. The main lobe of this spectrum is 2 Hz wide. The larger the symbol rate is, the larger the bandwidth of the signal.

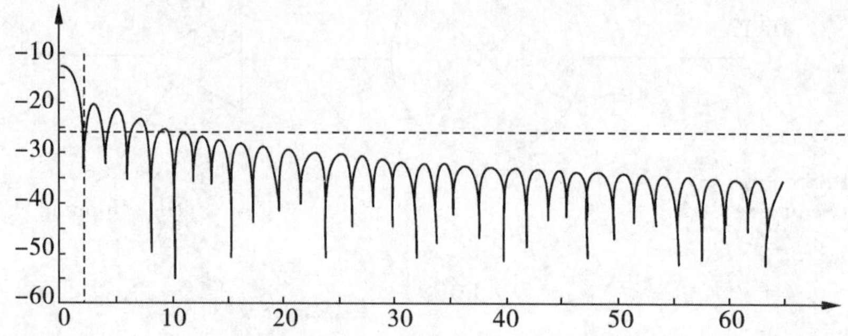

Figure 1–10 Spectrum of a binary signal with a data rate of 2 bps

An additional binary sequence (Figure 1–11) has a data rate 8 times larger than that of the sequence shown in figure 1–9.

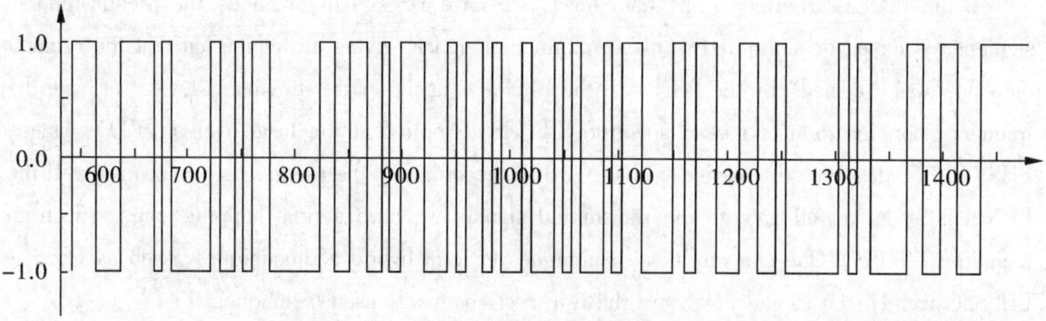

Figure 1–11 A new binary sequence used to modulate the information sequence

Instead of modulating with a sinusoid, we modulate sequence 1 with this new binary sequence, which we will call the code sequence for sequence 1. The resulting signal is shown in figure 1–12.

Since the bit rate has increased, we can assume that the spectrum of this sequence will have a larger main lobe.

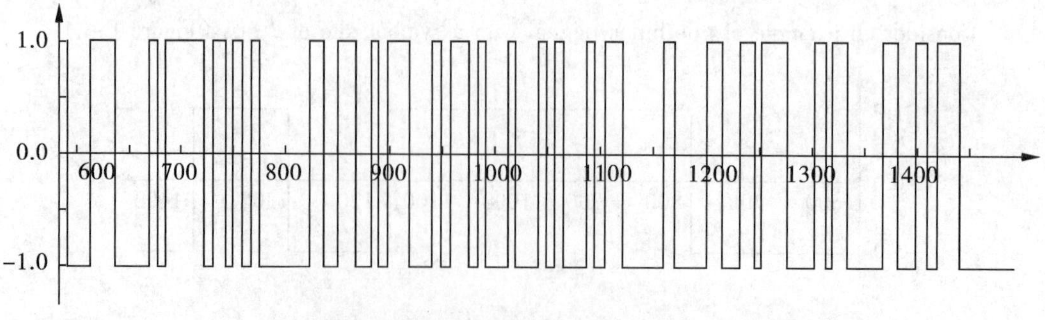

Figure 1–12 Each bit of sequence 1 is replaced by the code sequence

The spectrum of this signal has now spread over a larger bandwidth. The main lobe bandwidth is 16 Hz instead of 2 Hz. The process of multiplying the information sequence with the code sequence has caused the information sequence to inherit the spectrum of the code sequence (also called the spreading sequence) (Figure 1-13).

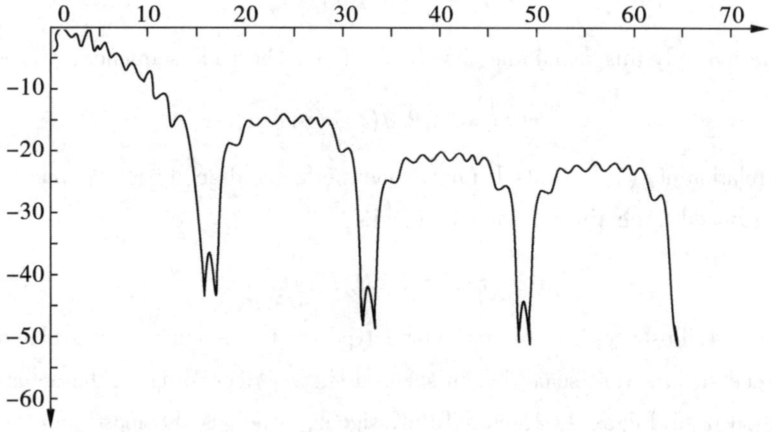

Figure 1-13 The spectrum of the spread signal is as wide as the code sequence

The spectrum has spread from 2 Hz to 16 Hz, i. e. , has increased by a factor of 8. This factor is called the spreading factor of the system, and this process can also be considered to be a form of binary modulation. Both the data signal and the modulating sequence in this case are binary signals.

If the original signal is $d(t)$ of power P_s and the code sequence is given by $g(t)$, the resultant modulated signal is

$$s(t) = \sqrt{2P_s}\, d(t) g(t) \tag{1.6}$$

The multiplication of the data sequence with the spreading sequence is the first modulation. Then, the signal is multiplied by the carrier, which is the second modulation. The carrier here is analog.

$$s(t) = \sqrt{2P_s}\, d(t) g(t) \sin(2\pi f_c t) \tag{1.7}$$

On the receiver side, we again multiply this signal with the carrier to obtain:

$$r(t) = \sqrt{2P_s}\, d(t) g(t)\ \sin^2(2\pi f_c t) \tag{1.8}$$

By trigonometric identity, we have

$$\sin^2(2\pi f_c t) = 1\text{-}\cos(4\pi f_c t) \tag{1.9}$$

We then obtain

$$r(t) = \sqrt{2P_s}\,d(t)g(t)[1 - \underline{\cos(4\pi f_c t]})\tag{1.10}$$

where the underlined part is the double frequency extraneous term, which we would filter out, leaving only the signal.

$$r(t) = \sqrt{2P_s}\,d(t)g(t)\tag{1.11}$$

Now, we multiply this remaining signalwith $g(t)$, the code sequence, and we obtain

$$r(t) = \sqrt{2P_s}\,d(t)g(t)g(t)\tag{1.12}$$

The correlation of $g(t)$ with itself (only when perfectly aligned) is a certain scalar number that can be removed to obtain the original signal.

$$r(t) = \sqrt{2P_s}\,d(t)\tag{1.13}$$

Figure 1 – 14 illustrates the most common type of direct-sequence-modulated SS signal. Direct-sequence spectra vary somewhat in spectral shape, depending on the actual carrier and data modulation used. Figure 1 – 15 is a BPSK signal, which is the most common modulation type used in direct-sequence systems.

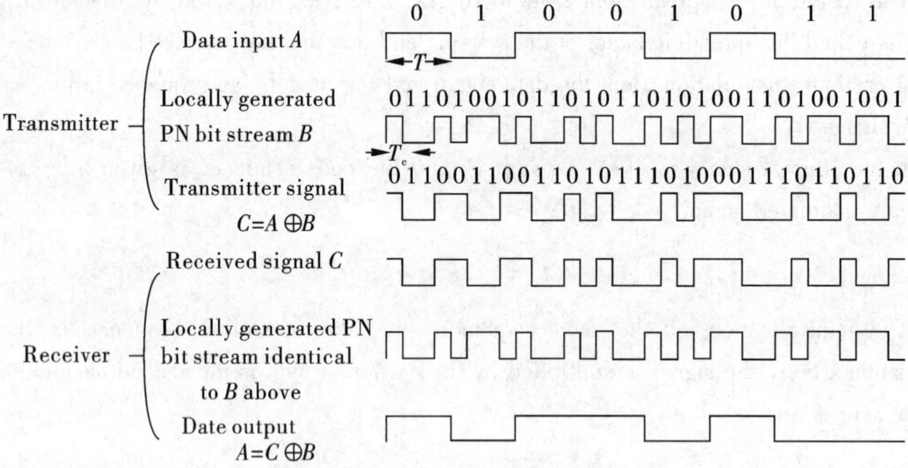

Figure 1–14 Example of direct-sequence spread spectrum

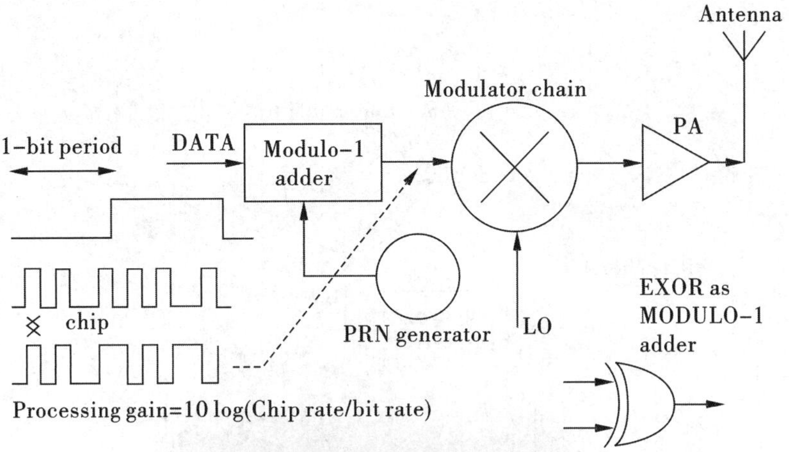

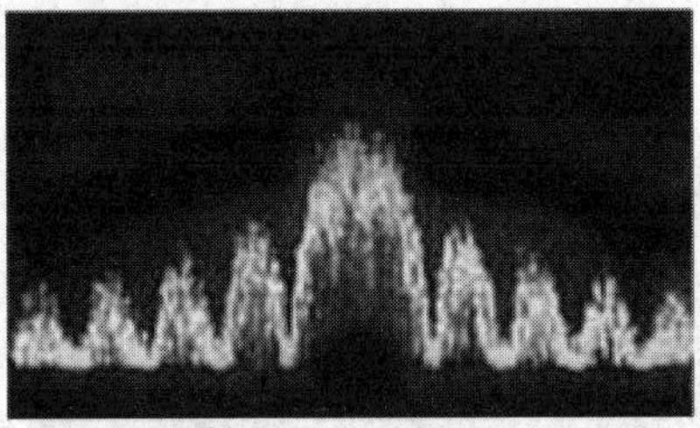

Figure 1-15　Spectrum-analyzer schematic of a direct-sequence spread-spectrum signal.

Note the original signal (non-spread) would occupy only half of the central lobe.

(2) **Frequency Hopping Spread Spectrum**

This method does exactly what its nameimplies, it causes the carrier to hop from frequency to frequency over a wide band according to a sequence defined by the PRN. The speed at which the hops are executed depends on the data rate of the original information, but one can distinguish between fast-frequency hopping (FFHSS) and low-frequency hopping (LFHSS). The latter method (the most common) allows several consecutive data bits to modulate the same frequency. On the other hand, FFHSS, is characterized by several hops within each data bit.

The transmitted spectrum of a frequency hopping signal is quite different from that of a direct-sequence system. Instead of a $[\sin(x)/x]^2$-shaped envelope, the frequency hopper output is flat over the band of frequencies used (Figure 1-16). The bandwidth of a frequency-hopping signal is simply N times the number of frequency slots available, where N is the bandwidth of each hop channel.

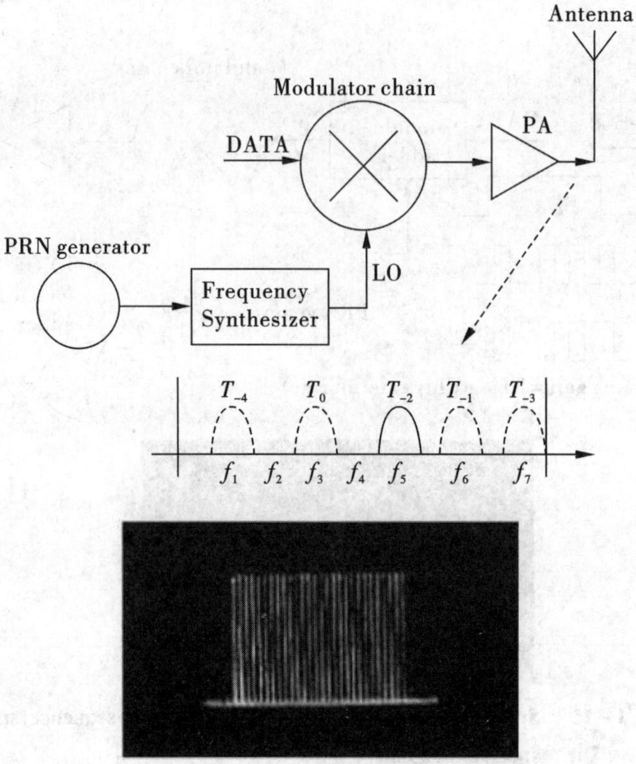

Figure 1-16 Frequency-hop spread-spectrum signal

(3) Time Hopping Spread Spectrum

THSS is illustrated in figure 1-17, where the on and off sequences applied to the PA are dictated by the PRN sequence.

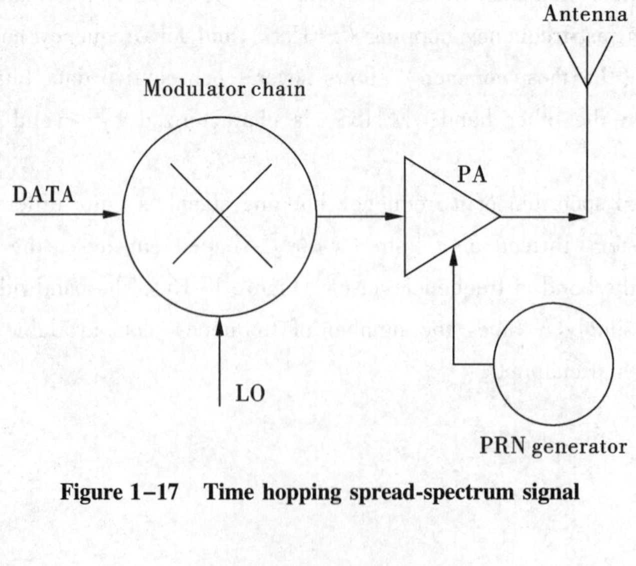

Figure 1-17 Time hopping spread-spectrum signal

1.2 Implementations

A complete SS communication link requires various advanced and up-to-date technologies and disciplines: RF antenna, powerful and efficient PA, low-noise, highly linear LNA, compact transceivers, high-resolution analog-to-digital converters and digital-to-analog converters, rapid low-power digital signal processing (DSP), etc. Although designers and manufacturers compete, they are also combining their effort to implement SS systems. The most difficult area is the receiver path, especially at the de-spreading level for DSSS, because the receiver must be able to recognize the message and synchronize with it in real time. The operation of code recognition is also called correlation. Because correlation is performed at the digital-format level, the tasks are mainly complex arithmetic calculations, including fast, highly parallel binary additions and multiplications. The most difficult aspect of today's receiver design is synchronization. More time, effort, research, and money has gone toward developing and improving synchronization techniques than toward any other aspect of SS communications.

Several methods can solve the synchronization problem, and many of them require a large number of discrete components to implement. Perhaps the greatest breakthroughs have occurred in DSP and in application-specific integrated circuits (ASICs). DSP provides high-speed mathematical functions to analyze, synchronize, and de-correlate an SS signal after slicing it into many small parts. ASIC chips drivedown costs via very-large-scale integration (VLSI) technology and by the creation of generic building blocks suitable for any type of application.

Exercises

1 Explanation of glossary

1-1 DSSS

1-2 FHSS

1-3 THSS

2 Choice question

2-1 Typical applications for the resulting short-range data transceivers are (   ).

A. GPS　　B. W-LAN

C. GSM notice board　　D. Bluetooth

2-2 (   ) are wideband technologies.

A. W-CDMA　　B. UMTS

C. CDMA　　D. TDMA

3 Short answer question

3-1 Introduce the basic principles of SS technology briefly.

3-2 Resume the advantages of spread-spectrum technology.

3-3 In the Gaussian white noise interference channel, the signal transmission bandwidth is 16 kHz, S/N is 4 dB, calculate channel capacity C.

3-4 Discuss the differences among DSSS, FHSS and THSS.

Reference

[1] MAXIM Inc. An Introduction to Direct-Sequence Spread-Spectrum Communications. Feb. 18, 2003 [2018-08-27]. https://people.cs.clemson.edu/~westall/851/spread-spectrum.pdf.

[2] LANGTON C. Intuitive Guide to Principles of Communications. 2002 [2018-08-27]. http://www.complextoreal.com/wp-content/uploads/2013/01/CDMA.pdf

Chapter 2 Multi-access communications

2.1 Multi-access channel

The idea of using a communication channel to enable several transmitters to send information simultaneously dates to Thomas A. Edison's 1873 invention of the diplex. This revolutionary system enabled the simultaneous transmission of two telegraphic messages in the same direction through the same wire. One message was encoded by changes in polarity, the other by changes in absolute value[1].

Currently, numerous examples of multi-access communication in which several transmitters share a common channel exist, including mobile telephones transmitting to a base station, ground station communication with a satellite, a bus with multiple taps, local area networks, packet-radio networks, and interactive cable television networks. A common feature of these communication channels is that the receiver obtains the superposition of the signals sent by the active transmitters (Figure 2-1).

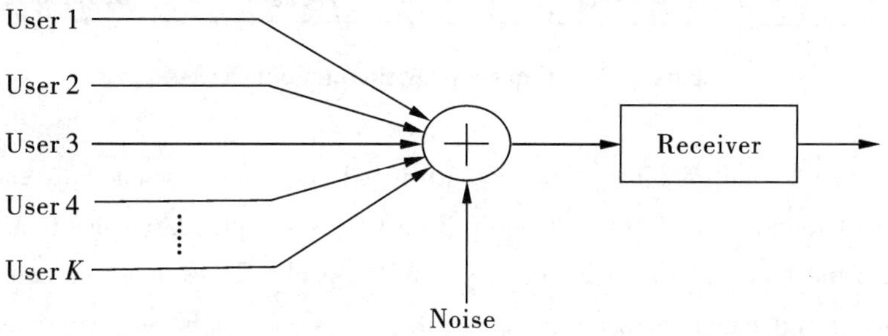

Figure 2-1 Multi-access communication

Oftentimes, the superposition of signals sent by different transmitters occurs unintentionally owing to non-ideal effects, for example, crosstalk in telephony and any time the same radio-frequency band is used simultaneously by distant transmitters, as in cellular telephony, radio/television broadcasting, and wireless local loops. Although the terms multiplexing and multi-access are sometimes used interchangeably, multi-access usually refers to situations where the message sources are not collocated and/or operate autonomously. The message sources in a multi-access channel are referred to as users[2].

The multi-access communication scenario depicted in figure 2-1 encompasses not only the case of a common receiver for all users, but the case of several receivers, each of which is interested in the information sent by one user. Multi-access communication is sometimes referred to as multipoint-to-point communication. The same information is delivered to all recipients, for example, in radio and television broadcasting and cable television. At the other extreme, the messages transmitted to different recipients are independent, for example, a base station transmitting to mobile units. The latter scenario falls conceptually within the multi-access channel model; in that case, the receiver is interested in only one of the information sources transmitted by the base station.

2.2 FDMA and TDMA

The advent of radio-frequency modulation in the early twentieth century enabled several radio transmissions to coexist in time and space without mutual interference by using different carrier frequencies. The same idea was used in long-distance wire telephony. Frequency-division multiplexing or frequency-division multiple access (FDMA) assigns a different carrier frequency to each user so that the resulting spectra do not overlap (Figure 2-2). Band-pass filtering enables separate demodulation of each channel.

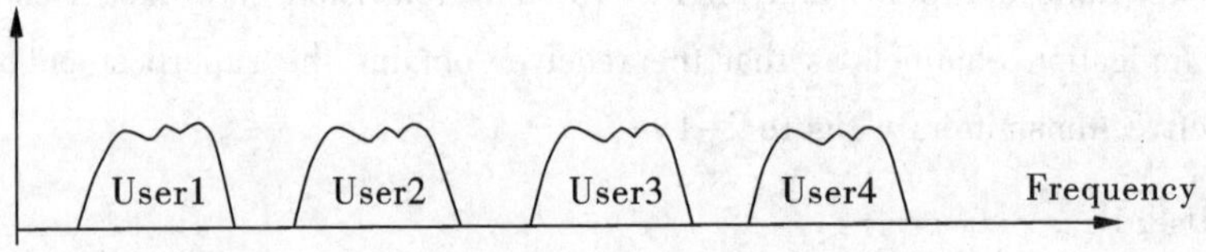

Figure 2-2 Frequency-division multiple access

In time-division multiplexing, time is partitioned into slots assigned to each incoming digital stream in round-robin fashion (Figure 2-3). De-multiplexing is conducted by simply switching on to the received signal at the appropriate epochs. Time divisions can be used not only to multiplex collocated message sources but also by geographically separated users who have the ability to maintain time synchronization, in what is commonly referred to as time-division multiple access (TDMA). Note that FDMA allows completely uncoordinated transmissions in the time domain: no time synchronization among users is required. This advantage is not shared by TDMA, where all transmitters and receivers must have access to a common clock.

Figure 2–3 Time-division multiple access with guard times between slots

The important feature of FDMA and TDMA is that, for all conceptual purposes, the various users are operating in separate noninterfering channels. In terms of the signal-space language of digital communications, the multi-access techniques operate by ensuring that the signals transmitted by the various users are mutually orthogonal. Channel or receiver non-ideal effects may require the insertion of guard times in TDMA (Figure 2–3) and spectral guard bands in FDMA (Figure 2–2) to avoid co-channel interference.

Why would it make sense to consider multi-access techniques that do not adhere to the principle of dividing the channel into independent noninterfering sub-channels? One reason is that noninterfering multi-access strategies may waste channel resources when the number of potential users is much greater than the number of simultaneously active users at any given time. For example, consider wireless telephony; if each subscriber were assigned a fixed RF channel, only a tiny fraction of the spectrum would be utilized at any given time. Analogously, in TDMA, most of the time slots would be empty at any given time.

How is it possible to assign the channel resources to users in dynamic rather than static fashion as above? At the expense of some increase in complexity, one possibility is tocreate a separate reservation channel where the users who want to use the channel notify the receiver, which then, partitions the original channel using TDMA or FDMA among active users only. This process presupposes a separate feedback channel that notifies every user of the time or frequency slot where it is allowed to transmit. However, the reservation channel is a multi-access channel, and we still have to cope with the same issue as before, namely, how to partition the resources of that channel dynamically.

FDMA is a good candidate for applications such as cordless telephones. In particular, the simple signal processing makes it a good choice for inexpensive implementation in a cordless environment.

On the other hand, in cellular applications, FDMA is inappropriate because of the lack of "built-in" diversity and the potential for severeintracell interference between base stations. A further complication arises from the difficulty of performing hand-overs if base stations are not tightly synchronized.

For personal communication systems, the decision is not as obvious. Depending on whether the envisioned personal communication system application more closely resembles more a cordless private branch exchange (PBX) than a cellular system, FDMA may be an appropriate choice. We will see below that it is probably better to opt for a combined TDMA/

FDMA or a CDMA-based system to avoid the pitfalls of pure FDMA systems while maintaining moderate equipment complexity.

2.3 Random multi-access

Random multi-access communication is one of the approaches to dynamic channel sharing. When a user has a message to transmit, the user transmits the message as if they were the sole user of the channel. If no other user is transmitting simultaneously, then the message is received successfully. However, the users are uncoordinated, and the possibility always exists that the message will interfere (in time and frequency) with another transmission. In such a case, it is typically assumed in random multi-access communication that the receiver cannot reliably demodulate several simultaneous messages. The only alternative remaining is to notify the transmitters that a collision has occurred and thus their messages have to be retransmitted. Collisions would reoccur forever if the transmitters involved were to retransmit immediately upon notification of a collision. To overcome this problem, users wait a random period of time before retransmitting. The main distinguishing feature among the existing random communications systems is the algorithm used by the transmitters to determine the retransmission delay for each colliding packet.

The first random multi-access communication system was the Aloha system proposed for a radio channel in 1969. Some coaxial-cable local area networks, typified by the widely used Ethernet, employ a "polite" version of Aloha, called carrier-sense multiple access, where users listen to the channel before transmitting so as not to collide with an ongoing transmission. In general, random multi-access communications are best suited for every bursty channels, in which it is unlikely that more than one user will be transmitting simultaneously. The main theoretical advances in this area occurred in the 1970s through the mid-1980s. Polling is another multi-access strategy to avoid simultaneous transmissions, in which the receiver asks every transmitter that shares a common channel whether it has anything to transmit.

2.4 CDMA

The channel-sharing approaches discussed so far are based on the philosophy of letting no more than one transmission occupy a given time-frequency slot. Whenever this condition is violated in random-access communication, the receiver is unable to recover any of the colliding transmissions. As we remarked before, reception free from inter-channel interference is a consequence of the use of orthogonal signaling. It is important to realize that this can be accomplished by signals that overlap in both time and frequency. For example, consider that the time-limited signals x_1 and x_2 in figure 2-4 overlap in both the time and frequency domains (Figure 2–5). Their cross-correlation or inner product is zero:

$$< x_1, x_2 > = \int_0^T x_1(t)x_2(t)\,dt = 0 \tag{2.1}$$

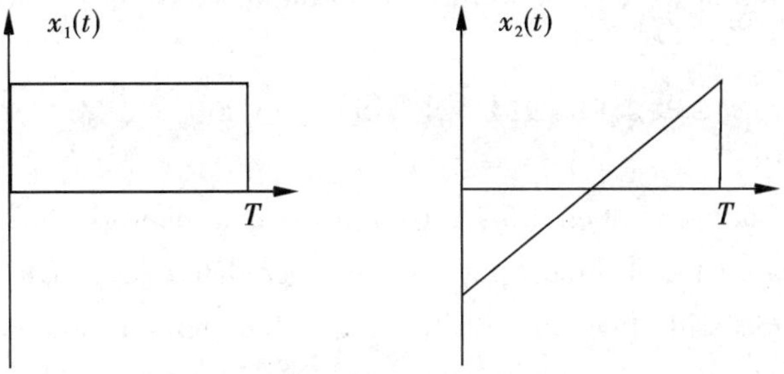

Figure 2-4 Orthogonal signals assigned to two users

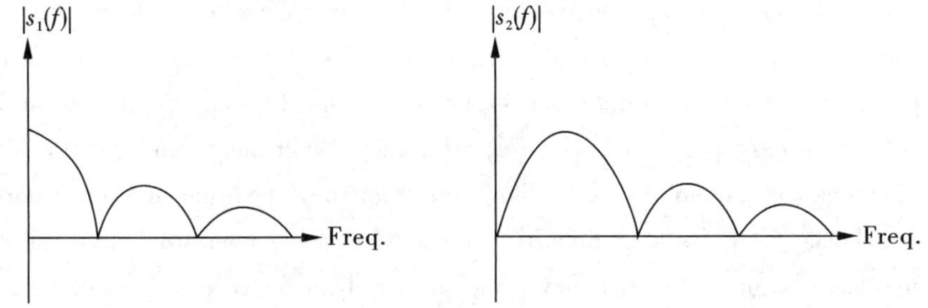

Figure 2-5 Fourier transforms (magnitude $|s_1(f)|>0$, $|s_2(f)|>0$) of the waveforms in Figure 2-4

Any receiver will actually observe the sum of both signals embedded in additive background noise. A good strategy in that case is matched filtering, in which the received waveform is separately correlated in every bit period with both $x_1(t)$ and $x_2(t)$, followed by respective comparisons of the correlator outputs to zero thresholds. The output of the correlator is affected by the background noise. However, due to the orthogonality condition (2.1) and assumed synchronism, the output of the correlator for user 1 is not affected by the signal of user 2, regardless of the relative strengths of the signals. We can conclude that although the signals transmitted by both users overlap in both time and frequency, the BER of this system will be the same as if the transmissions occurred in separate channels.

We have just seen a very simple example of CDMA. All users or base stations (BSs) operate in the same waveband. Each user is assigned a unique code sequence or signature sequence that enables the user to spread information across the assigned frequency band. The special case of orthogonal CDMA in which the signature waveforms do not overlap in the time domain corresponds to digital TDMA. Digital TDMA and FDMA systems can be seen as special cases of orthogonal CDMA, where the signature waveforms are non-overlapping in the time domain and the frequency domain, respectively. At the receiver, the signals from various users are separated by the cross-correlation of the received signal with each of the possible signature sequences of users. Carefully designed spread sequences with small cross-correlations can help to decrease the inherent problem of co-channel interference, which is usually called multiple access interference (MAI) within CDMA. This brings us to the objective of multiuser detection.

2.5 Multiple access techniques for 5G

The upcoming fifth generation (5G) mobile cellular networks are required to provide significant increases in network throughput and cell-edge data rates, high energy efficiency and low latency, compared with the currently deployed long-term evolution (LTE) and LTE-advanced networks. To satisfy these demands of 5G networks, innovative technologies for radio air-interface and radio access network are highly important in physical layer (PHY) design. Recently, non-orthogonal multiple access has attracted increasing research interest from both academia and industry as a potential radio access technique. A few examples include multiuser shared access, sparse code multiple access, resource spread multiple access and pattern division multiple access proposed by ZTE, Huawei, Qualcomm, and DTmobile. In the meantime, multicarrier technologies that divide the frequency spectrum into many narrow sub-channels, such as filter bank multicarrier (FBMC) and generalized frequency-division multiplexing, have become attractive new concepts for dynamic access spectrum management and cognitive radio applications.

Exercise

1 Explanation of glossary

1-1 FDMA

1-2 TDMA

1-3 CDMA

2 Choice question

2-1 Baud rate is (   )

A. the reciprocal of the shortest signalling element

B. the reciprocal of the longest signalling element

C. the reciprocal of the longest signalling element minus the shortest signalling element

D. always the same as the bit rate

2-2 The 4th generation of mobile communication allows the data transmission (   ).

A. 10 kbps ~ 100 kbps   B. 100 Mbps ~ 200 Mbps

C. 300 kbps ~ 30 Mbps   D. 500 Mbps ~ 5 Gbps

2-3 China's 3G communications standard is ( ).

A. TDMA B. CDMA

C. FDMA D. TD-SCDMA

3 Short answer question

3-1 How should we share our resources so that as many users as possible can communicate simultaneously?

3-2 What is the MAI problem in the CDMA?

Reference

[1] HAYKIN S. Communication systems. 4th ed. New York: John Wiley & Sons, 2000.

[2] VERDU S. Multiuser detection. Cambridge: Cambridge university press, 1998.

Chapter 3 Channel model

3.1 Basic concepts

A channel refers to the medium between the transmitting antenna and the receiving antenna, as shown in figure 3-1.

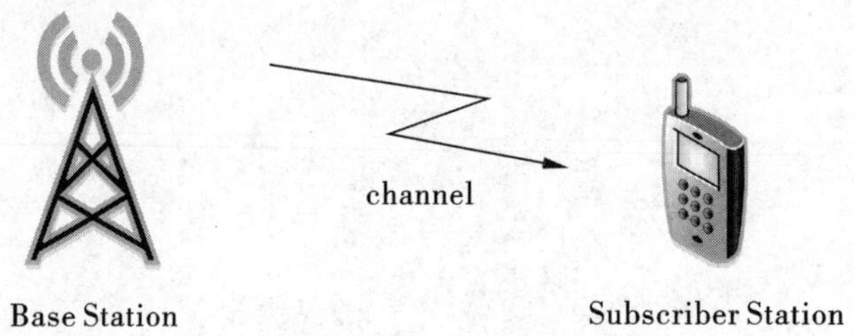

Figure 3-1 Channel

The characteristics of a wireless signal change as it travels from the transmitter antenna to the receiver antenna. These characteristics depend upon the distance between the two antennas, the paths taken by the signal, and the environment (buildings and other objects) around the path. The profile of the received signal can be obtained from that of the transmitted signal if we have a model of the medium between the two. This model of the medium is called the channel model.

The power profile of the received signal can be obtained by convolving the power profile of the transmitted signal with the impulse response of the channel. Convolution in the time domain is equivalent to multiplication in the frequency domain. Therefore, after propagation through the channel H , the transmitted signal x becomes y :

$$y(f) = H(f)x(f) + n(f) \tag{3.1}$$

where $H(f)$ is the channel response, and $n(f)$ is the noise. Note that x, y, H *and* n are all functions of the signal frequency f.

The three key components of the channel response are path loss, shadowing, and multipath, as explained below.

3.1.1 Path loss

The simplest channel is the free space line-of-sight (LOS) channel with no objects between the receiver and the transmitter or around the path between them. In this simple case, the transmitted signal attenuates because the energy is spread spherically around the transmitting antenna. For this LOS channel, the received power is given by:

$$P_r = P_t \left[\frac{\sqrt{G_l}\lambda}{4\pi d}\right]^2 \tag{3.2}$$

where P_t is the transmitted power, G_l is the product of the transmit and receive antenna field radiation patterns, λ is the wavelength, and d is the distance. Theoretically, the power falls off in proportion to the square of the distance. In practice, the power falls off more quickly, typically as the 3rd or 4th power of distance.

The presence of ground causes some of the waves to reflect and reach the transmitter. These reflected waves may sometimes have a phase shift of 180° and may reduce the net received power. A simple two-ray approximation for path loss can be shown to be:

$$P_r = P_t \frac{G_t G_r h_t^2 h_r^2}{d^4} \tag{3.3}$$

Here, h_t and h_r are the antenna heights of the transmitter and receiver, respectively. Note that there are three major differences from the previous formula. First, the antenna heights have an effect; second, the wavelength is absent; and third, the exponent on the distance is 4. A general empirical formula for path loss is:

$$P_r = P_t P_0 \left(\frac{d_0}{d}\right)^{\alpha} \tag{3.4}$$

where P_0 is the power at distance d_0, and α is the path loss exponent. The path loss is given by:

$$P_{\mathrm{loss}}(d)_{\mathrm{dB}} = \overline{P_{\mathrm{loss}}}(d_0) + 10\alpha\log\left(\frac{d}{d_0}\right) \tag{3.5}$$

Here, $\overline{P_{\mathrm{loss}}}(d_0)$ is the mean path loss in dB at distance d_0.

3.1.2 Shadowing

If there are any objects (such buildings or trees) along the path of the signal, some part of the transmitted signal is lost through absorption, reflection, scattering and diffraction. If the base antenna were a light source, the middle building would cast a shadow on the subscriber antenna, hence, this effect is called shadowing(Figure 3-2).

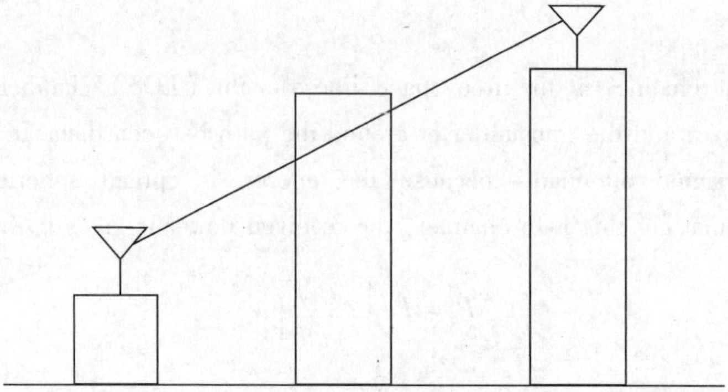

Figure 3-2 Shadowing

The net path loss becomes:

$$P_{\text{loss}}(d)_{\text{dB}} = \overline{P_{\text{loss}}}(d_0) + 10\alpha\log\left(\frac{d}{d_0}\right) + \chi \tag{3.6}$$

Here, χ is a normally (Gaussian) distributed random variable (in dB) with standard deviation σ, χ represents the effect of shadowing. As a result of shadowing, power received at points that are at the same distance d from the transmitter may be different and have a lognormal distribution.

3.1.3 Multipath

Objects located around the path of the wireless signal reflect the signal. Some of these reflected waves are also received at the receiver. Since each of these reflected signals takes a different path, it has a different amplitude and phase (Figure 3-3).

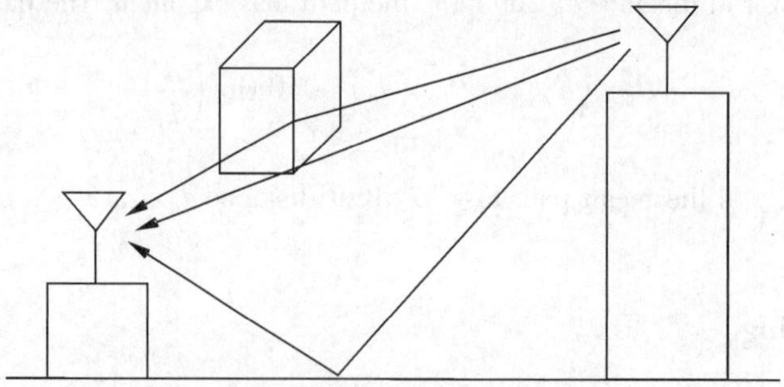

Figure 3-3 Multipaths

Depending upon the phase, these multiple signals may result in increased or decreased received power at the receiver. Even a slight change in position may result in a significant difference in the phases of the signals and therefore in the total received power. The three components of the channel response are shown clearly in figure 3 – 4. The thick dashed line represents the path loss. The lognormal shadowing changes the total loss to that shown by the thin dashed line. The multipath scenario results in the variations shown by the solid thick line. Note that the signal strength variation due to multipath changes at distances in the range of the signal wavelength.

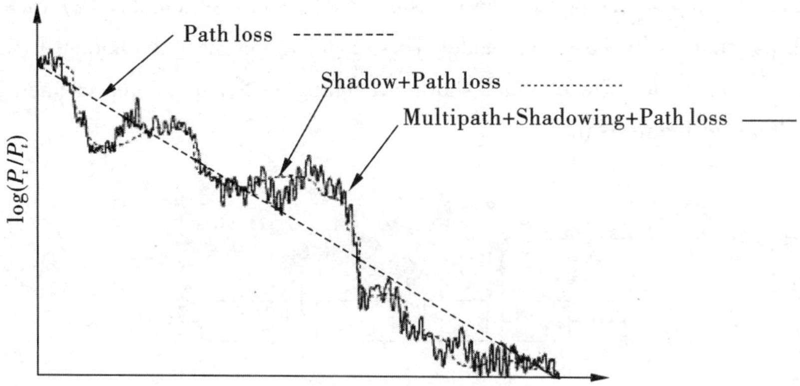

Figure 3-4 Path loss, shadowing and multipath

Since the paths have different lengths, a single impulse sent from the transmitter will result in multiple copies being received at different times, as shown in figure 3-5.

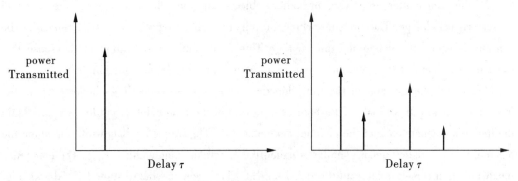

Figure 3-5 Multipath power delay profile

The maximum delay after which the received signal becomes negligible is called the maximum delay spread τ_{max}. A large τ_{max} indicates a highly dispersive channel. The root-mean-square (rms) value of the delay spread τ_{rms} is often used instead of the maximum.

3.1.4 Tapped delay line model

One way to represent the impulse response of a multipath channel is by a discrete number of impulses as follows:

$$h(t,\tau) = \sum_{i=1}^{M} c_i(t)\delta(\tau-\tau_i) \tag{3.7}$$

Note that the impulse response h varies with time t. The coefficients $c_i(t)$ vary with time. There are M coefficients in the above model. This model represents the channel by a delay line with M taps. For example, the channel shown in figure 3-5 can be represented by a 4-tap model, as shown in figure 3-6.

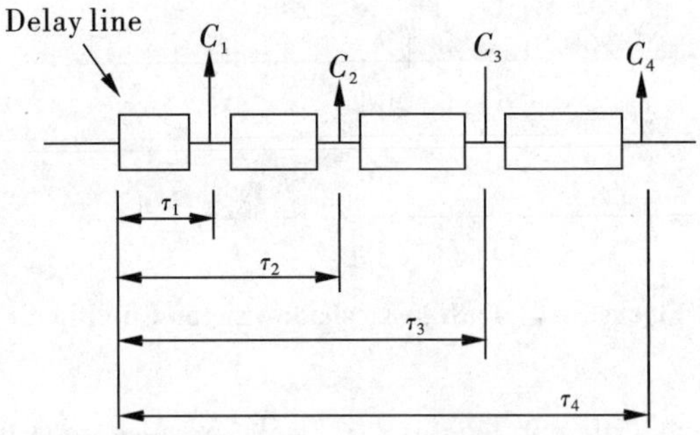

Figure 3-6 Tapped delay line model

If the transmitter, receiver, or other objects in the channel move, the channel characteristics change. The time for which the channel characteristics can be assumed to be constant is called the coherence time (T_c). This is a simplistic definition in the sense that exact measurement of the coherence time requires the autocorrelation function.

For every phenomenon in the time domain, there is a corresponding phenomenon in the frequency domain. If we look at the Fourier transform of the power delay profile, we can obtain the frequency dependence of the channel characteristics. The frequency bandwidth for which the channel characteristics remain similar is called the coherence bandwidth (B_c). The coherence bandwidth is inversely related to the delay spread. The larger the delay spread is, the smaller coherence bandwidth, and the channel is said to become more frequency selective.

3.1.5 Doppler spread

The power delay profile gives the statistical power distribution of the channel over time for a signal transmitted for just an instant. Similarly, the Doppler power spectrum gives the statistical power distribution of the channel for a signal transmitted at just one frequency f.

Whereas the power delay profile is the result of multipaths, the Doppler spectrum is caused by motion of the intermediate objects in the channel. The Doppler power spectrum is nonzero for $(f - f_D, f + f_D)$, where f_D is the maximum Doppler spread (B_D).

The coherence time and Doppler spread are inversely related:

$$\text{Coherence Time} \approx \frac{1}{\text{Doppler Spread}} \tag{3.8}$$

Thus, if the transmitter, receiver or intermediate objects moverapidly, the Doppler spread is large and the coherence time is small, i. e., the channel changes quickly.

3.2 Wireless channel

The term "wireless" is a generic word that indicates that electromagnetic waves or RF transmit information through a part of or the whole communication path. Wireless communication can transfer information over both short distances and long distances. Furthermore, it is commonly employed in telecommunication with an impractical or impossible use of wires. The users of next-generation wireless communication require higher information transmission rates and better transmission quality; however, wireless resources are correspondingly limited. Therefore, it is highly significant to analyze the properties of the wireless modes of wireless fading channel models.

The wireless modes include point-to-point communication, point-to-multipoint communication, multipoint-to-multipoint communication, and broadcasting. Figure 3-7 shows several wireless modes. Wireless devices, such as a cellphone, a wireless router, a laptop with Wi-Fi, and amplitude modulation or frequency modulation radio, utilize the electromagnetic spectrum from 9 kHz to 300 GHz in wireless communication. However, these frequencies are considered to be a public resource by most countries, and different ranges can be used for different purposes. For example, the frequency band of a common GSM cellphone is 900 MHz or 1800 MHz. The regular frequency band for a pilot to communicate with the control tower of an airport is also approximately 900 MHz. If a passenger is using a cellphone during takeoff or landing, the passenger's phone call may interfere with the pilot's communication with the control tower. Therefore, the efficient utilization of the limited channel resources is highly meaningful.

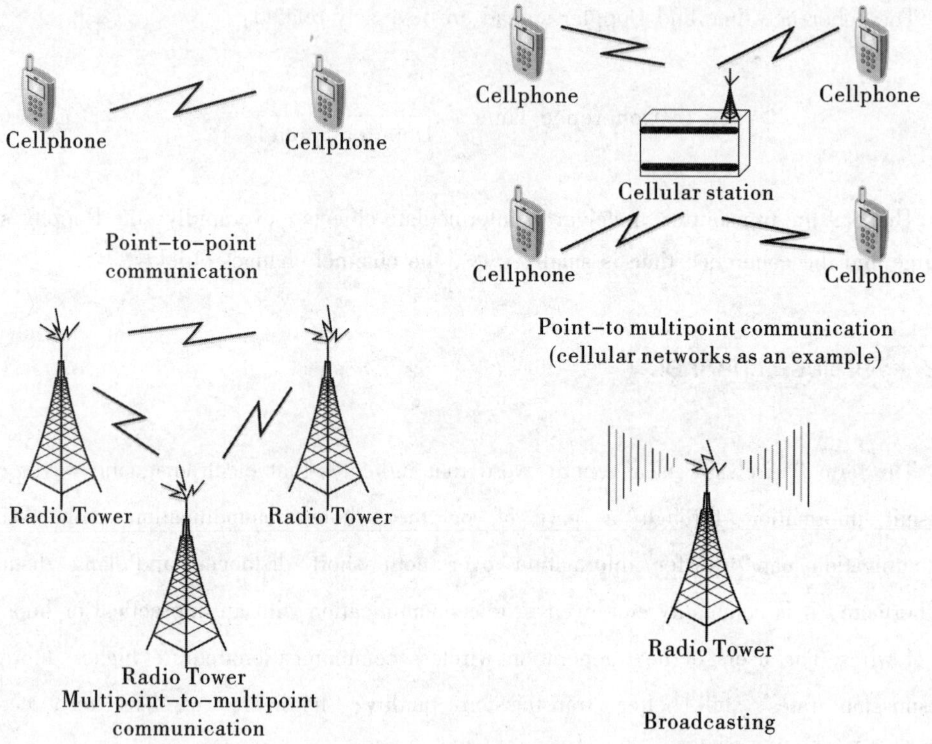

Figure 3-7 Wireless modes

In broadband wireless connections, the symbol rate must be increased further, which leads to a frequency-selective channel. The recent development of the communication technology has built upon the great interest in multi-antenna systems as an effective technique to combat fading and to reduce the effect of channel interference.

3.2.1 Fading

Fading is a type of attenuation that can occur when transmitting modulated signals over some propagation media. In wireless communication, fading may be due to either multipath propagation or shadowing.

Figure 3-8 shows that obstacles such as buildings, clouds, tress and planes can reflect, scatter, and diffract the transmitted waves. The waves transmitted waves by non-LOS paths are usually obstructed by those obstacles. Therefore, the signals received through the LOS are generally stronger than those from none-LOS paths.

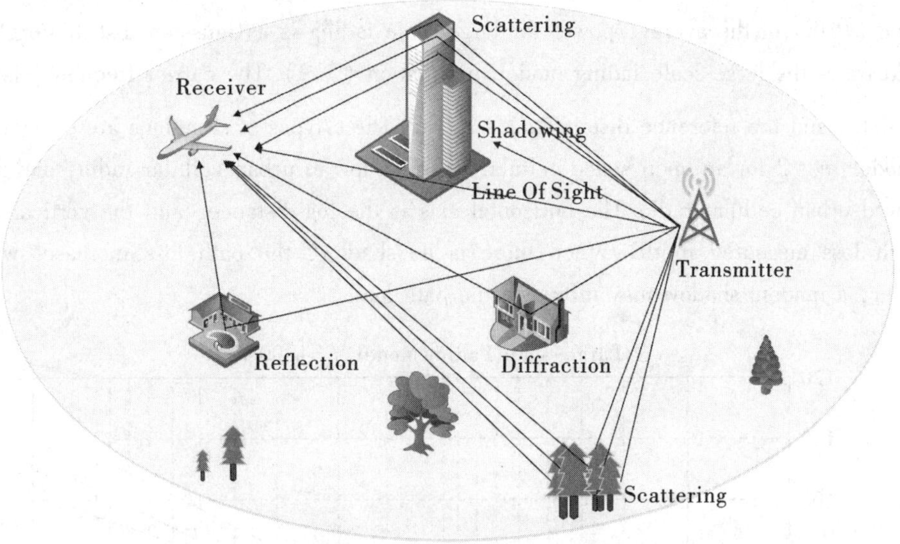

Figure 3–8 Fading and multipath

Different time, frequency or location could change the fading, and common fading channel models include two aspects: large-scale fading and small-scale fading. Large-scale fading, which is also called attenuation or path loss, is related to large distances or the time-average characteristics of transmitted signals. By contrast, small-scale fading, is the fast fluctuation of the amplitude or power of the transmitted signals corresponding to short distances or short time intervals. Attenuation and several fading channels containing flat fading, frequency-selective fading, slow fading, fast fading, Rayleigh fading, and Rician fading should be considered.

3.2.2 Large-scalefading model

Large-scale fading, or attenuation, represents path loss by propagation, including reflection, scattering and diffraction, in an open space environment. The average power at a distance $\hat{d}$ from the transmitter is given as

$$P_{\hat{d}} = \beta\,(\hat{d}/\hat{d}_0)^{-v} P_t \tag{3.9}$$

where β is a large-scale fading parameter that depends on the antenna gain, frequency, wavelength, and other factors; $\hat{d}_0$ is the reference distance; P_t is the average transmitted power; and v is the path loss exponent.

The path loss exponent is usually equal to 2 in an open space environment and is greater than 2 in an environment with obstacles, as shown in figure 3–8. In a practical or empirical situation, such as the cellphone in figure 3–8, the $P_{\hat{d}}$ of the phone may not be the same in different locations at the same distance from the transmitting tower or the base station because these obstacles can randomly influence path loss through shadowing. The probability density function (PDF) of the average power for large-scale fading is a Gaussian distribution. Figure 3-9 illustrates the large-scale fading model in equation (3.9). The carrier frequency is set to 1800 MHz, and the reference distance $\hat{d}_0$ is 100 m. Three types of situations are considered in this model: v is 2 for an open space cellular radio, 3 for an urban cellular radio, and 5 for a shadowed urban cellular radio. The horizontal axis is the log-distance, and the vertical axis is the path loss measured in dB. When there is no shadow, the path loss increases with v. However, a random shadow may influence the path loss.

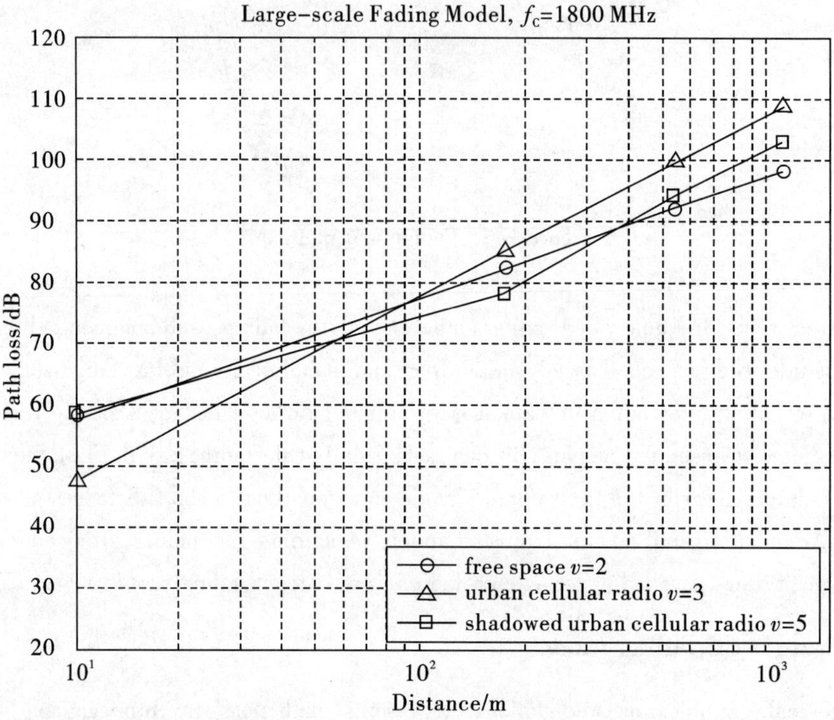

Figure 3-9 Large-scale fading model

3.2.3 Small-scalefading model

Small-scale fading, or simply fading, denotes a fast fluctuation in the power of the transmitted signals when the receivers are moved within a relatively small area. Fading is caused by the multipath waves of the transmitted signals, and these waves reach the receivers at moderately different times. As mentioned before, scattering, diffraction, and reflection can generate multipath waves, and this process is known as multipath fading. Depending on the relation between the signal parameters (such as bandwidth, symbol period, etc) and the channel parameters (such as rms delay spread and Doppler spread), different transmitted signal will undergo different of fading. While multipath delay spread leads to time dispersion and frequency selective fading, Doppler spread leads to frequency dispersion and time selective fading. Figure 3-10 shows a tree of the four different types of fading.

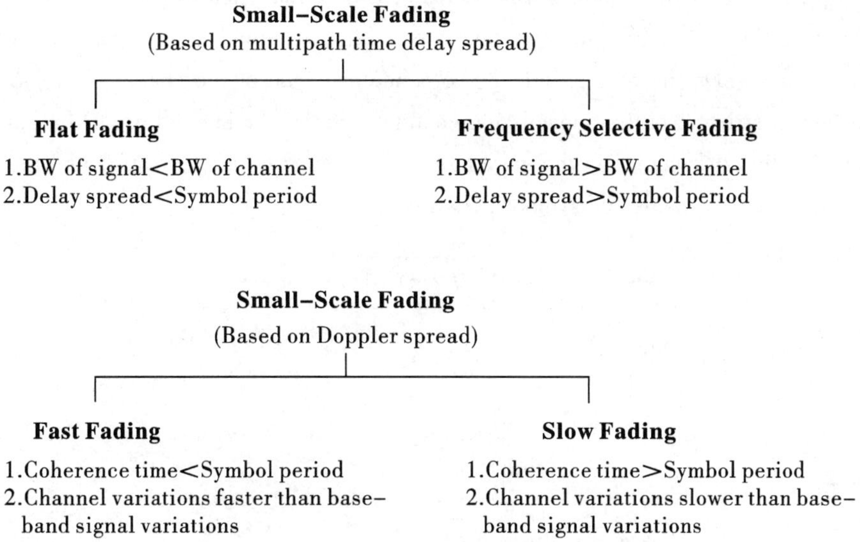

Figure 3-10 Types of small-scale fading.

To research the behavior of different fading channels, a general channel model is depicted in figure 3-11. In the time and frequency domains, the source signals, channel response, and received signals are $s(t)$ and $S(f)$, $h(t)$ and $H(f)$, and $r(t)$ and $R(f)$, respectively.

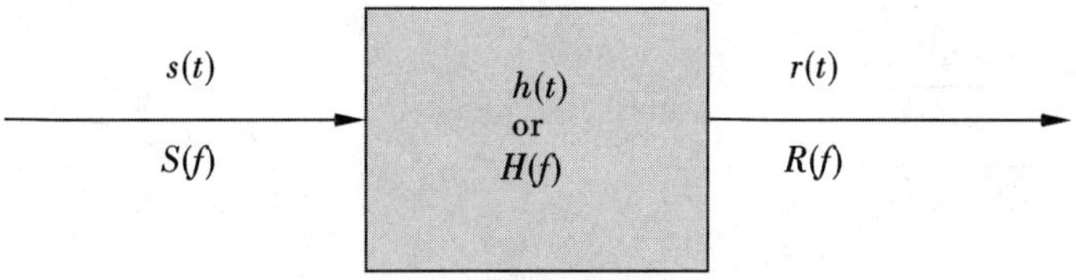

Figure 3-11 General channel model

Figure 3-12(a)(b) shows two types of wireless channels: a flat fading channel and a frequency-selective fading channel. In a flat fading channel, the coherence bandwidth (B_c) of the channel is larger than the bandwidth of the source signal (B_s). Here, the coherence represents the minimum frequency or time required for the magnitude change of the channel. On the other hand, under the opposite conditions, with a larger bandwidth of the source signal and a smaller coherence channel bandwidth in the frequency domain, as shown in figure 3-12(b), the received signals will be distorted. If we consider $S(f)$ as the symbol bandwidth in the frequency domain, it is assumed that intersymbol interference (ISI) exists. Therefore, the channel bandwidth $H(f)$ in the frequency domain should be equal to or larger than the coherence channel bandwidth in the frequency domain to avoid ISI.

Compared with the delay restriction of the transmission channel, slow fading will occur if the coherence time of this channel is not small. The shadowing mentioned above can lead to slow fading. The rate of the channel will be much slower than that of the transmitted signal. Then, the variations of amplitude and phase can be considered static over one or several bandwidth intervals. On the other hand, if the coherence time of the channel is relatively less than the delay restriction of the transmission channel, the fast fading will occur. In this case, it is worth noting that the change in the amplitude and phase is no longer static, and the channel response alters rapidly inside the symbol period.

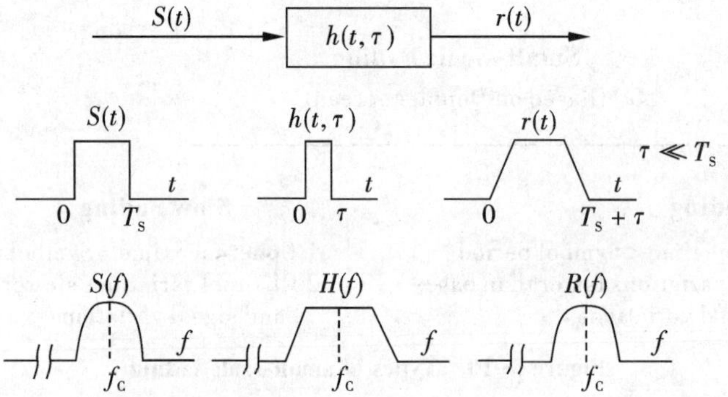

(a) Flat fading channel characteristics

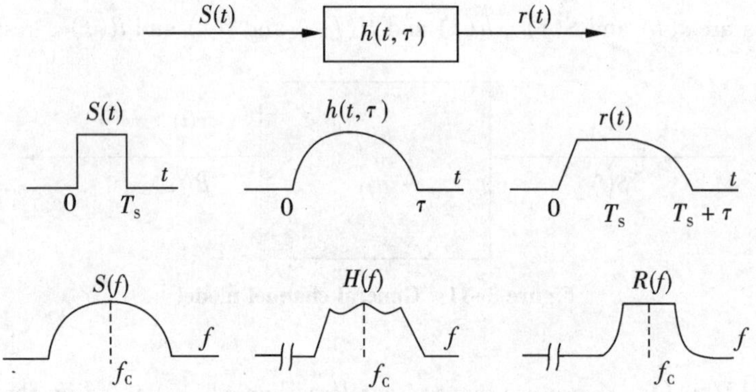

(b) Frequency-selective fading channel characteristics

Figure 3 – 12 Flat fading channel and frequency-selective fading channel characteristics

Overall, the flat or frequency-selective fading channel and the slow or fast fading channel are modeled by a linear time-varying impulse response. The signal undergoes flat fading if $B_s < B_c$ and $T_s > \sigma_\tau$ and the signal undergoes frequency-selective fading if $B_s > B_c$ and $T_s < \sigma_\tau$. Alternatively, the channel is frequency selective if $T_s \leqslant 10\sigma_\tau$, where T_s is the reciprocal bandwidth (e. g., symbol period) and σ_τ is the rms delay spread. However, the nature or practical environment of the multipath channels makes it impossible to maintain such an ideal impulse response. Thus, statistical models are considered to explore the performance of the received signals, and the most essential models include the Rayleigh fading channel model and the Rician fading channel model.

There are mainly two cases in a flat fading channel. If there is no LOS path between the transmitter and the receiver, it is a Rayleigh fading model. Otherwise, it is a Rician fading channel. The PDFs of a Rayleigh random variable and a Rician random variable are given by equation (3.10) and equation (3.11), respectively:

$$f_{\text{rayleigh}}(r) = \frac{r}{\sigma^2}\exp\left(\frac{-r^2}{2\sigma^2}\right), r \geqslant 0 \tag{3.10}$$

$$f_{\text{rician}}(r) = \frac{r}{\sigma^2}\exp\left(\frac{-(r^2 + D^2)}{2\sigma^2}\right) I_0\left(\frac{D_r}{\sigma^2}\right), r \geqslant 0, D \geqslant 0 \tag{3.11}$$

where r is the magnitude of the received signal, σ^2 is the average power, D indicates the peak amplitude of the LOS signal, $I(\cdot)$ is the modified Bessel function of the first kind, and $I_0(\cdot)$ is the zero-order of $I(\cdot)$. The Rician distribution approaches the Rayleigh distribution when there is no LOS path in the channel, i. e., when D is very close to zero.

Exercise

1 Explanation of glossary

1-1 Shadowing

1-2 Multipath

1-3 Fading

2 Choice question

2-1 The three key components of the channel response are ( ).

A. Path loss B. Multipath

C. Shadowing D. Doppler spread

2-2 In a flat fading channel, if there is no LOS path between the transmitter and the receiver, it is a ( ) model.

A. Rayleigh fading B. Log-normal fading

C. Rician fading D. Nakagami-m

3 Short answer question

3-1 Describe the tapped delay line model and what it is used for.

3-2 What is multipath fading? How it will affect the signal transmission?

3-5 Describe the flat fading, frequency-selective fading, slow fading, fast fading in your words.

Reference

[1] JAIN R. Channel Models – A Tutorial. Feb. 21, 2007[2018-08-27]. http://www.cse.wustl.edu/~jain/cse574-08/ftp/channel_model_tutorial.pdf

[2] RAPPAPORT T S. Wireless communications: principles and practice. vol. 2. New Jersey: prentice hall PTR, 1996.

[3] YU D. Multiuser Detection in Multiple Input Multiple Output Orthogonal Frequency Division Multiplexing Systems by Blind Signal Separation Techniques. Florida International University, 2012[2018-08-25]. http://digitalcommons.fiu.edu/cgi/viewcontent.cgi?article=1737&context=e.

Chapter 4 Mathematical models and hardware implementations

4.1 Solving normal systems of equations

A wide variety of signal processing applications require the linear least squares (LS) problem[1] to be solved in real time; among these are adaptive antenna arrays[2], multiuser detection[3], MIMO detection[4], echo cancellation[5], equalization[6], and system identification[1]. The linear LS problem is equivalent to the solution of a system of linear equations, also referred to as normal equations[7]

$$\boldsymbol{A}\boldsymbol{x} = \boldsymbol{b} \tag{4.1}$$

where $\boldsymbol{A}$ is an $N \times N$ symmetric positive definite matrix, and $\boldsymbol{x}$ and $\boldsymbol{b}$ are both $N \times 1$ vectors. Matrix $\boldsymbol{A}$ and vector $\boldsymbol{b}$ are known, whereas vector $\boldsymbol{x}$ must be estimated. An exact solution is generally defined as

$$\boldsymbol{x} = \boldsymbol{A}^{-1}\boldsymbol{b} \tag{4.2}$$

where $\boldsymbol{A}^{-1}$ is the inverse of matrix $\boldsymbol{A}$. The computational load of matrix inversion is related to the size of the matrix N and is generally regarded as an operation of a complexity $O(N^3)$[8]. Standard mathematical calculation software, such as MATLAB, uses one of a variety of techniques from the LAPACK library[9] to solve this problem. From a numerical perspective, the best approach to matrix inversion is to not do it explicitly but instead, where possible, to solve an applicable system of equations[10-12]. Consequently, for real-time solutions, techniques that solve systems of equations are the most suitable approach to the LS problem. Many efficient methods exist and can be categorized under two main headings: direct methods and iterative methods. Direct methods compute an exact solution of the system of equations through a finite number of pre-specified operations[10]. By contrast, iterative methods produce a sequence of successively better approximations of the optimal solution[10].

These methods can also be used to calculate the inverse of matrix $\boldsymbol{A}^{-1}$. Let $\boldsymbol{AX} = \boldsymbol{I}$, where $\boldsymbol{I}$ is an $N \times N$ identity matrix and $\boldsymbol{X} = \boldsymbol{A}^{-1}$ is an $N \times N$ matrix to be calculated; we obtain N systems of equations

$$\boldsymbol{A}\boldsymbol{X}_{:,n} = \boldsymbol{I}_{:,n} \quad n = 1, 2, \cdots, N \tag{4.3}$$

We can obtain $\boldsymbol{A}^{-1}$ by solving these N systems of equations. Direct and iterative algorithms for solving linear systems are presented in the following two subsections.

4.1.1 Direct methods

Direct methods, such as Gaussian elimination, $\boldsymbol{LU}$ factorization, Cholesky decomposition, and QRD, obtain an exact solution of the system of equations (4.1) after a finite sequence of pre-specified operations[10]. The key idea of most direct methods is to reduce the general system of equations to an upper or lower triangular form that has the same solution as the original equations. The equations can be solved easily by back or forward substitution, which often provides a highly accurate solution[8].

The basic idea of Gaussian elimination is to modify the original equations (4.1) to obtain an equivalent triangular system by taking appropriate linear combinations of the original equations (4.1)[8]. Specifically, the process systematically applies row operations to transform the system of equations (4.1) to an upper triangular system $\boldsymbol{Ux} = \boldsymbol{y}$, where $\boldsymbol{U}$ is an $N \times N$ upper triangular matrix and $\boldsymbol{y}$ is an $N \times 1$ vector. Then, the upper triangular system $\boldsymbol{Ux} = \boldsymbol{y}$ can be solved easily through back substitution. The complexity of the Gaussian elimination method is as high as $2N^3/3$ operations[8], including multiplications, divisions and additions. In addition to high complexity, the main disadvantage of the Gaussian elimination method is that the right-hand vector $\boldsymbol{b}$ of (4.1) is involved in the elimination process and has to be known in advance for the elimination step to proceed[8].

$\boldsymbol{LU}$ factorization (or $\boldsymbol{LU}$ decomposition) can be viewed as a high-level algebraic description of Gaussian elimination[8] that decomposes matrix $\boldsymbol{A}$ of the system of equations (4.1) into a product, $\boldsymbol{A} = \boldsymbol{LU}$, where $\boldsymbol{L}$ is a unit lower triangular matrix with all the main diagonal elements equal to one, and $\boldsymbol{U}$ is an upper triangular matrix[8]. Therefore, the solution vector $\boldsymbol{x}$ can be obtained by solving a lower triangular system $\boldsymbol{Ly} = \boldsymbol{b}$ by forward substitution and an upper triangular system $\boldsymbol{Ux} = \boldsymbol{y}$ by back substitution sequentially[8]. Compared with the Gaussian elimination technique, the benefit of $\boldsymbol{LU}$ decomposition is that the matrix modification (or decomposition) step can be executed independently of the right-side vector $\boldsymbol{b}$ [8]. Thus, when we have solved the system of equations (4.1), we can solve additional systems with the same left-side matrix $\boldsymbol{A}$ without further matrix decomposition[10]. Therefore, the complexity of solving equivalent systems with the same left-side matrix is significantly reduced. This property of $\boldsymbol{LU}$ decomposition has a great meaning in practice, which makes $\boldsymbol{LU}$ decomposition the method of choice in many applications[13]. However, $\boldsymbol{LU}$ decomposition is complicated, requiring as many as $2N^3/3$ operations[8], including multiplication, division and addition.

As the coefficient matrix $\boldsymbol{A}$ in the normal equations (4.1) is symmetric and positive definite, the efficient Cholesky decomposition method can be used. Cholesky decomposition is closely related to Gaussian elimination[10]. It decomposes the positive definite coefficient matrix $\boldsymbol{A}$ in exactly one way into a product $\boldsymbol{A} = \boldsymbol{U}^{\mathrm{T}}\boldsymbol{U}$, where $\boldsymbol{U}$ is an upper triangular matrix with all the main diagonal elements positive[10]. Consequently, the system of equations (4. 1) can be rewritten as $\boldsymbol{U}^{\mathrm{T}}\boldsymbol{U}\boldsymbol{x} = \boldsymbol{b}$. Let $\boldsymbol{y} = \boldsymbol{U}\boldsymbol{x}$, we obtain a lower triangular system $\boldsymbol{U}^{\mathrm{T}}\boldsymbol{y} = \boldsymbol{b}$. Therefore, the solution vector $\boldsymbol{x}$ can be easily obtained by solving a low triangular system $\boldsymbol{U}^{\mathrm{T}}\boldsymbol{y} = \boldsymbol{b}$ and an upper triangular system $\boldsymbol{U}\boldsymbol{x} = \boldsymbol{y}$ sequentially through forward and back substitution, respectively. Compared with Gaussian elimination, Cholesky decomposition has the advantage that it requires half of the number of operations and half of the memory space[8]. Furthermore, Cholesky decomposition is numerically stable[8] because positive definite matrix $\boldsymbol{A}$ is nonsingular. However, the complexity of Cholesky decomposition is as high as $N^3/3$ operations, including multiplication and division[8]. Therefore, Cholesky decomposition is still too complicated for real-time hardware implementations, especially if the system size N is large[8].

QRD is known for its numerical stability[8] and is widely used in many applications. It solves the system of equations (4. 1) in the same way as $\boldsymbol{LU}$ decomposition[10]; it transforms the coefficient matrix $\boldsymbol{A}$ as $\boldsymbol{A} = \boldsymbol{Q}\boldsymbol{R}$, where $\boldsymbol{Q}$ is an orthogonal matrix and $\boldsymbol{R}$ is an upper triangular matrix[8]. The orthogonal matrices have the property of $\boldsymbol{Q}\boldsymbol{Q}^{\mathrm{T}} = \boldsymbol{Q}^{\mathrm{T}}\boldsymbol{Q} = \boldsymbol{I}$ and $\boldsymbol{Q}^{-1} = \boldsymbol{Q}^{\mathrm{T}}$ [8], where $\boldsymbol{I}$ is an $N \times N$ identity matrix. Therefore, system (4. 1) can be transformed into an upper triangular system $\boldsymbol{R}\boldsymbol{x} = \boldsymbol{Q}^{\mathrm{T}}\boldsymbol{b}$, and the solution vector $\boldsymbol{x}$ can be obtained easily through back substitution.

QRD is equivalent to computing an orthogonal basis for a set of vectors[8]. Several methods can be used to compute the QRD, such as Householder reflections and Givens rotations[8]. Reflections and rotations are computationally attractive as they are easily constructed and can be used to introduce zeros in a vector by choosing an appropriate rotation angle or refection plane[8]. Householder reflections are extremely useful for introducing zeros to annihilate all elements (except the first) of a vector[8]. By contrast, Givens rotations introduce zeros to a vector more selectively, including the first element[8]. Therefore, Givens rotations are usually the transformation of choice[8].

By using Givens rotations, QRD is inherently well suited to hardware implementation, exploiting parallel processing and pipeline capabilities of a systolic array structure[14,15], which consists of an array of individual processing cells arranged as a triangular structure. Each individual processing cell in such an array has its own local memory and is connected to only its nearest cells[1]. The special architecture of the array makes regular streams of data be pipelined through the array in a highly rhythmic fashion. This simple and highly parallel systolic array enables simple data flow and high throughput with pipelining[4]. Therefore, it is well suited to implementing complex signal processing algorithms, particularly for real-time and high-data-bandwidth implementations[1]. However, the complexity of this triangular array architecture is highly related to the system size, the number of processing cells $(N^2 + N)/2$ grows dramatically with increasing matrix size N, which makes direct hardware design for the systolic array very expensive for most practical applications, e. g. , adaptive beamforming, thereby requiring multiple-chip rather than single-chip solutions[16]. Therefore, the triangular architecture is only feasible for matrices with small size[4].

Alternatively, some less complicated architecture arrays can be used to solve large matrices. In [17], a linear architecture systolic array for QRD is obtained through direct projection of the triangular architecture systolic array; each processing cell of the linear array corresponds to the processing cells of each row in the triangular array. This linear array reduces the number of processing cells to N. However, the processing cell of the linear array in [17] is much more complicated than that of the triangular array, as it is obtained by merging the functions of the diagonal and off-diagonal cells of the triangular array. Moreover, not all the processing cells are utilized 100%. Another type of linear architecture systolic array[18,16] is obtained by using the folding and mapping approach on the triangular systolic array. This type of linear array retains the local interconnections of the triangular systolic array and requires only M processing cells ($N = 2M + 1$ is the matrix size). These processing cells have complexity similar to that of the triangular array. Moreover, these processing cells are 100% utilized. In [15], another solution was proposed to avoid the triangular architecture array by combining similar processing cells of the triangular array, adding memory blocks and using a control logic to schedule the data movement between blocks. The combined architecture is less complicated at the expense of heavier latency compared to the linear architecture. Therefore, the linear architecture arrays[16,17] and the combined architecture[15] provide a balanced trade-off between performance and complexity[16]. They require more control logic and heavier latency, but the required number of processing cells is significantly reduced[4] compared to that of the triangular structure systolic array.

The classic Givens rotations contain square root, division and multiplication operations[15], which are expensive for hardware implementation. Abundant work has been performed on efficient systolic array implementation of QRD using Givens rotations on hardware, and these methods can be divided into three main types. The first type of QRD using Givens rotations is based on the Coordinate Rotation Digital Computer (CORDIC) algorithm. The CORDIC algorithm is an iterative technique for computing trigonometric functions, such as sine and cosine[15]. The CORDIC algorithm is simple as it requires only bit-shift and addition operations[15]. Therefore, CORDIC is suitable for fixed-point hardware implementation. However, due to the limited dynamic range of the fixed-point representation in CORDIC algorithms, the word-length requirement for the same accuracy is much greater than that in floating-point implementations[15]. Moreover, larger errors occur due to the many sub-rotations of the CORDIC algorithm[15].

The second type of QRD using Givens rotations is based on a square-root-free version of the Givens rotations or squared Givens rotations (SGR)[19], which eliminates the need for square-root operations and eliminates half of the multiplications[15]. The SGR method has several benefits compared to the classic Givens rotation and CORDIC technique. First, it is simple because it does not require square-root operations. However, its numerical accuracy is lower than that of the classic Givens rotations with square-root operations. Second, it is much faster than the CORDIC technique[15,4]. The SGR-based hardware design is approximately half as large and results in only approximately 66% latency compared to that of the CORDIC-based hardware design[15,4]. Furthermore, the SGR method can also be implemented using floating-point representation, which requires 80% of the number of bits required by the fixed-point representation of CORDIC algorithm for the same accuracy[8,20].

The third type of QRD using Givens rotations is based on logarithmic number systems (LNS) arithmetic. In LNS arithmetic, the square-root operations of the conventional number system become simple bit-shift operations, and the multiplication and division operations of the conventional number system become addition and subtraction operations, respectively[21]. However, the simple addition and subtraction operations in the conventional number system become much more costly in LNS arithmetic[21]. Therefore, although the Givens rotation operations using LNS arithmetic are multiplication-free, division-free and square-root-free, the addition operations make hardware implementation challenging. In addition, LNS-based QRD requires a number of converting operations to fit into the conventional number system design.

Traditionally, systolic array-based QRD is used for large systems[22,23]. For small matrices, there are alternative algorithms that are faster and more hardware efficient than QRD while still providing sufficient numerical stability[22,23]. One straightforward matrix-inversion method is the analytic approach[22]. For example, the inversion of a 2 × 2 matrix using the analytic approach is computed as follows[22]:

$$\boldsymbol{B}^{-1} = \begin{bmatrix} a & b \\ c & d \end{bmatrix}^{-1} = \frac{1}{ad-bc} = \begin{bmatrix} d & -b \\ -c & a \end{bmatrix} \tag{4.4}$$

For size matrices, the complexity of the analytic approach is significantly less than that of the QRD. Therefore, the analytic approach is efficient for inversions of small matrices, such as the 2 × 2 matrix in equation (4.4). However, the complexity of the analytic approach grows rapidly as the size N of the matrix increases, which makes it suitable for only small matrices. Moreover, direct analytic matrix inversion is sensitive to finite-length errors[22]. Even for 4 × 4 matrices, the direct analytic approach is unstable due to the large number of subtractions involved in the computation, which might introduce cancellation[22].

A method called blockwise analytic matrix inversion (BAMI) is proposed in [22] to compute the inversion of complex-valued matrices. BAMI partitions the matrix into four smaller matrices and then computes the inverse based on the computations of these smaller parts. For example, to compute a 4 × 4 matrix $\boldsymbol{B}$, it is first divided into four 2 × 2 submatrices[22]

$$\boldsymbol{B} = \begin{bmatrix} \boldsymbol{B}_1 & \boldsymbol{B}_2 \\ \boldsymbol{B}_3 & \boldsymbol{B}_4 \end{bmatrix} \tag{4.5}$$

Consequently, the inversion of matrix $\boldsymbol{B}$ can be computed via the inversion of these 2 × 2 matrices using the analytic method (4.4), i. e., [22]

$$B^{-1}=\begin{pmatrix} B_1^{-1}+B_1^{-1}B_2(B_4-B_3B_1^{-1}B_2)^{-1}B_3B_1^{-1} & -B_1^{-1}B_2(B_4-B_3B_1^{-1}B_2)^{-1} \\ -(B_4-B_3B_1^{-1}B_2)^{-1}B_3B_1^{-1} & (B_4-B_3B_1^{-1}B_2)^{-1} \end{pmatrix} \tag{4.6}$$

The BAMI approach is more stable than the direct analytic method, due to the fewer number of subtractions, and it requires fewer bits to maintain the precision[22]. Therefore, the BAMI method provides a good alternative to the classic QRD for solving small matrices, such as 4×4 matrices. However, for 2×2 and 3×3 matrices, the direct analytic method is preferred[22].

The Sherman-Morrison equation is a special case of the matrix-inversion lemma, allowing easy computation of the inverse of a series of matrices where two successive matrices differ only by a small perturbation[8]. The perturbation has to have the form of a rank-1 update, e. g., uv^{H}, where u and v are vectors of appropriate size[24]. Given A^{-1}, the Sherman-Morrison formula is expressed as[8]

$$(A^{-1}+uv^{H})^{-1}=A^{-1}-\frac{(A^{-1}uv^{H})A^{-1}}{1+v^{-1}A^{-1}u} \tag{4.7}$$

Consequently, for a series of matrices $A(i)=A(i-1)+u(i)u^{H}(i)$, e. g., autocorrelation matrices, where i is the time index and $u(i)$ is the input vector, $A^{-1}(i)$ can be computed easily by using the Sherman-Morrison formula[24], i. e.,

$$A^{-1}(i)=[A(i-1)+u(i)u^{H}(i)]^{-1}=A^{-1}(i-1)-\frac{A^{-1}(i-1)u^{H}(i)A^{-1}(i-1)}{1+u^{H}(i)A^{-1}(i-1)u(i)} \tag{4.8}$$

This approach to matrix inversion is widely used, e. g., in MIMO systems[25]. However, equation (4.8) requires a large number of multiplications and divisions, which makes the hardware implementation of this method difficult. In [24], the divisions in (4.8) are transformed into multiplications by introducing appropriate scaling, i. e.,

$$\begin{aligned}\widetilde{A}^{-1}(i)&=[\alpha(i-1)+u^{H}(i)\widetilde{A}^{-1}(i-1)u(i)]\left[\widetilde{A}(i-1)+\frac{u(i)u^{H}(i)}{\alpha(i-1)}\right]^{-1}\\ &=\widetilde{A}^{-1}(i-1)[\alpha(i-1)+u^{H}(i)\widetilde{A}^{-1}(i-1)u(i)]-[\widetilde{A}^{-1}(i-1)u(i)u^{H}(i)\widetilde{A}^{-1}(i-1)]\end{aligned} \tag{4.9}$$

where $\widetilde{A}^{-1}(i)=\alpha(i)\widetilde{A}^{-1}(i)$ and the scaling factor

$$\alpha(i)=\alpha(i-1)[\alpha(i-1)+u^{H}(i)\widetilde{A}^{-1}(i-1)u^{H}(i)] \tag{4.10}$$

with $\alpha(0)=1$. However, the modified Sherman-Morrison equation (4.9) remains complicated, with a complexity of $O(N^3)$ multiplications, which is expensive for the hardware design.

The direct methods, such as the Gaussian, Cholesky and QRD, have complexity of $O(N^3)$ operations, including division and multiplication operations. The modified Sherman-Morrison method requires approximately $O(N^2)$ multiplications. Therefore, direct methods are difficult in terms of real-time signal processing and hardware implementation. The direct methods compute an exact solution after a finite number of pre-specified operations[8] and output results only after executing all the prespecified operations. Therefore, if we stop early, the direct methods output no results[10]. Moreover, the direct methods may be prohibitively expensive when solving very large or sparse systems of linear equations[8].

4.1.2 Iterative methods

Iterative methods, which are efficient for both very large systems and very sparse systems, are an alternative to direct methods[10]. Iterative methods produce a sequence of successively better approximations $\boldsymbol{x}(k)$ (k is the iteration index), which ideally converge to the optimal solution[10] and involve the coefficient matrix $\boldsymbol{A}$ only in the context of the matrix-vector multiplication operations[8].

Solving the normal system of equations (4.1) can be formulated as minimizing the quadratic function[10]

$$f(\boldsymbol{x}) = \frac{1}{2}\boldsymbol{x}^{\mathrm{T}}\boldsymbol{A}\boldsymbol{x} - \boldsymbol{x}^{\mathrm{T}}\boldsymbol{b} \tag{4.11}$$

The minimum value of $f(\boldsymbol{x})$, obtained by setting $\boldsymbol{x} = \boldsymbol{A}^{-1}\boldsymbol{b}$, is $-\frac{1}{2}\boldsymbol{b}^{\mathrm{T}}\boldsymbol{A}^{-1}\boldsymbol{b}$, which is exactly the solution of the normal equations (4.1)[8]. Consequently, most iterative methods solve the normal equations (4.1) by minimizing the function $f(\boldsymbol{x})$ iteratively[8]. Each begins from an initial guess $\boldsymbol{x}^{(0)}$ and generates a sequence of iterates $\boldsymbol{x}^{(1)}, \boldsymbol{x}^{(2)}, \cdots$. In each step (or iteration), $\boldsymbol{x}^{(k+1)}$ is chosen as $f(\boldsymbol{x}^{(k+1)}) \leqslant f(\boldsymbol{x}^{(k)})$, and $f(\boldsymbol{x}^{(k+1)}) < f(\boldsymbol{x}^{(k)})$ is preferable[10]. Therefore, we approach the minimum value of $f(\boldsymbol{x})$ step by step. If we obtain $\boldsymbol{A}\boldsymbol{x}^{(k)} = \boldsymbol{b}$ or nearly so after some iterations, we can stop and accept $\boldsymbol{x}^{(k)}$ as the solution of the system (4.1).

The computation of the step from $\boldsymbol{x}^{(k)}$ to $\boldsymbol{x}^{(k+1)}$ has two components: ①choosing a direction vector $\boldsymbol{p}^{(k)}$ that indicates the direction of travel from $\boldsymbol{x}^{(k)}$ to $\boldsymbol{x}^{(k+1)}$; ②choosing a point on the line $\boldsymbol{x}^{(k)} + \alpha^{(k)}\boldsymbol{p}^{(k)}$ as $\boldsymbol{x}^{(k+1)}$, where $\alpha^{(k)}$ is the step size chosen to minimize $f(\boldsymbol{x}^{(k)} + \alpha^{(k)}\boldsymbol{p}^{(k)})$[10]. The process of choosing $\alpha^{(k)}$ is called the line search[10]. We want to choose an appropriate $\alpha^{(k)}$ to make $\boldsymbol{x}^{(k+1)} \leqslant \boldsymbol{x}^{(k)}$. One way to ensure this is to choose $\alpha^{(k)}$ to let

$$f(\boldsymbol{x}^{(k+1)}) = \min f(\boldsymbol{x}^{(k)} + \alpha^{(k)}\boldsymbol{p}^{(k)}) \tag{4.12}$$

This is called an exact line search; otherwise, the process is an inexact line search[10].

Two main types of iterative methods exist: nonstationary methods and stationary methods. Nonstationary methods, including the steepest descent method and the CG method, are a relatively recent development. They are usually complicated, but they can be highly effective[26]. Stationary iterative methods are older and simple but are usually not as effective as nonstationary methods[26]. Three stationary iterative methods are commonly used for linear systems: Jacobi, Gauss-Seidel and successive over-relaxation (SOR) methods.

One well-known iterative technique is the steepest descent method[8], which performs the exact line search in the direction of the negative gradient

$$\boldsymbol{p}^{(k)} = -\nabla f(\boldsymbol{x}^{(k)}) = \boldsymbol{b} - \boldsymbol{A}\boldsymbol{x}^{(k)} = \boldsymbol{r}^{(k)} \tag{4.13}$$

and we call $\boldsymbol{r}^{(k)}$ the residual vector of the solution $\boldsymbol{x}^{(k)}$. The step size is chosen as[8]

$$\alpha^{(k)} = \frac{(\boldsymbol{r}^{(k)})^{\mathrm{T}} \boldsymbol{r}^{(k)}}{(\boldsymbol{r}^{(k)})^{\mathrm{T}} \boldsymbol{A} \boldsymbol{r}^{(k)}} \tag{4.14}$$

The steepest descent method is easy to program, but it often converges slowly[10]. The main reason for its slow convergence is the time spent on minimizing $f(\boldsymbol{x}^{(k)})$ along parallel or nearly parallel search directions[10]. The complexity of the steepest descent method is high, requiring $O(N^2)$ multiplications, divisions and additions per iteration.

The CG algorithm is a simple variation of the steepest descent method that has a fast convergence speed. It is based on the idea that the speed of convergence to the optimal solution could be accelerated by minimizing $f(\boldsymbol{x}^{(k)})$ over the hyperplane that contains all previous search directions, i. e. ,

$$\boldsymbol{x}^{(k)} = \alpha^{(0)} \boldsymbol{p}^{(0)} + \alpha^{(1)} \boldsymbol{p}^{(1)} + \cdots \alpha^{(k-1)} \boldsymbol{p}^{(k-1)} \tag{4.15}$$

instead of minimizing $f(\boldsymbol{x}^{(k)})$ over just the line that points down the gradient[10], as in the steepest descent method. Due to its fast convergence, the CG method has been used for adaptive filtering for a long time (e. g. , see [27]-[30] and references therein). However, the complexity of the CG iteration is $O(N^2)$, including divisions, multiplications and additions, which is often too high for real-time signal processing.

The Jacobi method is perhaps the simplest iterative method[8]. It updates the next iterate $\boldsymbol{x}^{(k+1)}$ beginning with an initial guess $\boldsymbol{x}^{(0)}$ by solving each element of $\boldsymbol{x}$ in terms of[8]

$$\boldsymbol{x}_n^{(k+1)} = (b_n - \sum_{p \neq n} A_{n,p} x_p^{(k)}) / A_{n,n} \tag{4.16}$$

where $A_{n,p}$, b_n, and $\boldsymbol{x}_n$ are the (n,p)-th elements of the coefficient matrix $\boldsymbol{A}$ and the n-th elements of vectors $\boldsymbol{b}$ and $\boldsymbol{x}$, respectively. The Jacobi method has the advantage that all the elements of the correction $\boldsymbol{x}^{(k+1)}$ can be performed simultaneously because all elements of the new iterate $\boldsymbol{x}^{(k+1)}$ are independent of each other; therefore, the Jacobi method is inherently parallel[10]. On the other hand, the Jacobi method does not use the most recently available information to compute $\boldsymbol{x}_n^{(k+1)}$[8], as shown in (4.16). Therefore, the Jacobi method must store two copies of $\boldsymbol{x}$ because $\boldsymbol{x}^{(k)}$ can only be overwritten until the next iterate $\boldsymbol{x}^{(k+1)}$ is obtained[10]. If the system size N is large, each copy of $\boldsymbol{x}$ will occupy large memory space. Moreover, the Jacobi method requires nonzero diagonal elements of matrix $\boldsymbol{A}$ to avoid division by zero in (4.16), which can usually be achieved by permuting rows and columns if the condition is not already true. Furthermore, the Jacobi method does not always converge to the optimal solution[26]. However, the convergence of the Jacobi method is guaranteed under conditions that are often satisfied (e. g. , if matrix $\boldsymbol{A}$ is strictly diagonally dominant), even though the convergence speed may be very slow[26].

The Gauss-Seidel method is obtained by revising the Jacobi iteration to make it use the most current estimation of the solution $\boldsymbol{x}$ [10]; the Gauss-Seidel iterations are performed using each new component as soon as it has been computed rather than waiting until the next iteration[10], as in the Jacobi method. This feature gives the Gauss-Seidel method in terms of[10].

$$\boldsymbol{x}_n^{(k+1)} = \left(b_n - \sum_{p<n} A_{n,p}\boldsymbol{x}_p^{(k+1)} - \sum_{p>n} A_{n,p}\boldsymbol{x}_n^{(k)}\right) / A_{n,n} \tag{4.17}$$

Therefore, the Gauss-Seidel method can store each new element $\boldsymbol{x}_n^{(k+1)}$ immediately in the place of the old $\boldsymbol{x}_n^{(k)}$, saving memory space and making programming easier[26]. On the other hand, the Gauss-Seidel iterations can only be performed sequentially as each component of $\boldsymbol{x}^{(k+1)}$ depends only on previous ones; therefore, the Gauss-Seidel method is inherently sequential[10]. Moreover, the Gauss-Seidel method requires some conditions to guarantee its convergence; the conditions are somewhat weaker than those for the Jacobi method (e. g., if the matrix is symmetric and positive definite)[26]. The Gauss-Seidel method converges faster and requires only slightly more than half as many iterations as the Jacobi method to obtain the same accuracy[10]. Due to the explicit division in (4.17), the Gauss-Seidel method also requires nonzero diagonal elements of matrix $\boldsymbol{A}$. Gauss-Seidel iterations are widely used, such as in adaptive filtering[31, 32].

Relaxation is the process of correcting an equation by modifying one unknown[10]. The Jacobi method performs simultaneous parallel relaxation and the Gauss-Seidel method performs successive relaxation[10]. Over-relaxation is a technique to make a larger correction, rather than making only a correction for which the equation is satisfied exactly[10]. Over-relaxation can substantially accelerate the convergence[10]. SOR is achieved by applying extrapolation to the Gauss-Seidel technique. SOR successively takes the weighted average of the previous iterate and the current Gauss-Seidel iteration, i. e.,

$$\boldsymbol{x}^{(k+1)} = \omega\, \boldsymbol{x}_{GS}^{(k+1)} + (1-\omega)\, \boldsymbol{x}^{(k)} \tag{4.18}$$

where $\boldsymbol{x}_{GS}^{(k+1)}$ is the next iterate given by the Gauss-Seidel method, and $\omega > 1$ is the extrapolation factor[26]. The value of ω determines the acceleration of the convergence speed. If $\omega = 1$, SOR collapses to the Gauss-Seidel method. The parameter ω can also be chosen as $\omega < 1$, which amounts to under-relaxation, but this choice normally leads to slow convergence[10]. With the optimal value for ω, SOR can be an order of magnitude faster than the Gauss-Seidel method[26]. However, choosing the optimal ω is generally difficult, except for special classes of matrices[26]. As SOR is based on successive relaxation (Gauss-Seidel iterations), it also requires only a single copy of $\boldsymbol{x}$.

Compared with the direct methods, the iterative methods have several advantages: ①they require less memory than direct methods; ②they are faster than direct methods; ③they handle special structures (such as sparsity) in a simpler way[10]. Furthermore, the iterative techniques have the ability to exploit a good initial guess, which could reduce the number of iterations required to obtain the solution[10]. Although an infinite number of iterations might theoretically be required to converge to the optimal solution, iterative techniques have the ability to stop the solution process arbitrarily based on the required accuracy. By contrast, the direct methods do not have the ability to exploit an initial guess and simply execute a predetermined sequence of operations to obtain the solution[10].

Iterative methods are complex. Nonstationary iterative methods, such as the steepest descent method and the CG method presented above, are highly effective, with a complexity of $O(N^2)$ operations per iteration. The stationary iterative methods, such as the Jacobi, Gauss-Seidel and SOR algorithms, are less complicated, with a complexity of $O(N)$ operations per ite-ration, at the expense of lower efficiency. The iterative methods require division and multiplication operations, making them expensive for real-time signal processing and hardware design.

The DCD algorithm[3] is a nonstationary iterative method based on stationary coordinate descent techniques. Implementation of the DCD algorithm is simple because it does not require multiplication or division operations. In each iteration, the DCD algorithm requires only $O(N)$ additions or $O(1)$ additions. Therefore, the DCD algorithm is suitable for hardware realization.

4.2 Hardware reference implementations

The processes for solving linear systems of equations and matrix inversion have long been considered to be too difficult to implement within real-time systems. Consequently, hardware reference designs have started to appear only relatively recently. Based on our knowledge, most of the related hardware designs are based on the QRD using Givens rotations.

Karkooti *et al.* [15] implemented a 4×4 floating-point complex-valued matrix-inversion core on the Xilinx Virtex-4 XC4VLX200 FPGA chip based on the QRD algorithm via SGR. The design uses 21-bit data format: 14 bits for the mantissa, 6 bits for the exponent of the floating-point number and 1 sign bit. The design was implemented using the Xilinx System Generator tool[33], calling on the Xilinx Core Generator[34] to implement a floating-point divider block. To make the design fit on a single chip, a combined systolic array architecture was implemented. Internally, the design consists of one diagonal cell, one off-diagonal internal cell and a back-substitution block. The design also requires block RAMs and a control unit to schedule the movement of data between these blocks. Compared to the triangular architecture systolic array, the combined architecture makes the design less complicated at the expense of heavier latency. The area usage of this design is approximately 9117 logic slices and 22 DSP48 (also known as "XtremeDSP") blocks[35] with a latency of 933 clock cycles (777 cycles for QRD and 156 cycles for back substitution). The Virtex-4's DSP48 block is a configurable multiply accumulate block based on an 18-bit × 18-bit hardware multiplier[35]. The design can be extended to other sized matrices with a slight modification of the control unit and the RAM size.

Edman *et al.* [17] implemented a 4×4 fixed-point complex-valued matrix-inversion core on the Xilinx Virtex-II FPGA chip using QRD via SGR. The design is based on a linear architecture systolic array obtained through direct projection of the triangular architecture systolic array. This linear array requires only $2N$ processing cells for QRD and back substitutions. However, the processing cells are complicated and not all of them are of 100% utilized. The most complicated complex-valued divider was realized using 9 multipliers, 3 adders and a look-up table, with 5 pipelining stages. The implementation with 19-bit fixed-point data format consumes 86% of the Virtex-II chip area with a latency of 175 cycles. Unfortunately, the paper does not detail the FPGA chip model and the area usage in terms of the logic slices, RAMs and multipliers.

Liu *et al.* [16] implemented a floating-point QRD array processor based on SGR for adaptive beamforming on a TSMC (Taiwan Semiconductor Manufacturing Company) 0.13 micron chip. The design is based on an N-cell linear architecture systolic array obtained by using the folding and mapping approach on the triangular systolic array. The processing cell has similar complexity to that of the cell in the triangular systolic array and has a utilization of 100%. By using parameterized arithmetic processors, the design provides an elegant and direct approach to create a generic core for implementing the QR array processor. For the case of a 41-element antenna system in which data are represented by a 14-bit mantissa and 5-bit exponent, the linear array QR processor comprises 21 processing cells and utilizes 1060 K gates, corresponding to a maximum clock rate of 150 MHz. The authors did not implement their linear array with an FPGA chip. However, as the basic operations of SGR are complex, we could conclude that the FPGA implementation of this linear array is complicated and it is not possible to implement a large system, such as a 41-element system, on a single FPGA chip.

Myllyla *et al.* [4] implemented a fixed-point complex-valued minimum mean squared error (MMSE) detector for a MIMO OFDM system for both the 2×2 and 4×4 cases on the Xilinx Virtex-II XC2V6000 FPGA chip. The matrix operation of this design is based on the systolic array using CORDIC-based QRD and SGR-based QRD. A fast and parallel triangular structure of the systolic array is considered for 2×2 antenna systems and a less complicated linear architecture with easy scalability and time sharing processing cells is considered for 4×4 systems. The CORDIC-based design is implemented using VHDL, whereas the SGR-based design is implemented using the System Generator. For 2×2 and 4×4 systems, the CORDIC-based QRD design, in which data are represented in a 16-bit fixed-point data format, requires 11910 and 16805 slices, 6 and 101 block RAMs, 20 and 44 18-bit $\times$ 18-bit embedded multipliers, with 685 and 3000 cycles latency, respectively. The SGR-based QRD implemented for only the 2×2 system, requires 6305 slices, 8 block RAMs and 59 18-bit $\times$ 18-bit embedded multipliers with a latency of 415 cycles while using 19-bit fixed-point data format. Clearly, the CORDIC-based design requires more slices and fewer multipliers than those of the SGR-based design because the SGR is based on normal arithmetic operations, whereas CORDIC is based on multiplication-free and division-free rotation operations[4].

Many commercial QR intellectual property (IP) cores use the CORDIC algorithm. Altera has published a CORDIC-based QRD-RLS design[36][37] using their CORDIC IP core[38] to support applications such as smart antenna beamforming[2], Worldwide Interoperability for Microwave Access (WiMAX)[39], channel estimation and equalization of 3G wireless communications[40]. Altera's CORDIC IP block has a deeply pipelined parallel architecture, enabling speeds over 250 MHz on their Stratix FPGAs. In [36], [37], they explore a number of different degrees of parallelism of CORDIC blocks to perform matrix decomposition for 64 input vectors for a 9-element antenna with 16-bit data on an Altera Stratix FPGA. The required logic resources of the CORDIC cores can be as low as 2600 logic elements (equivalent to 1300 Xilinx slices[41][42]). When the CORDIC cores run at 150 MHz, the design obtains an update rate of 5 kHz and the processing latency is approximately 29700 cycles. The logic resources of other modules are not given. An embedded NIOS processor is used to perform the back substitution with a latency of approximately 12000 cycles for 9×9 matrix. However, the QRD and back substitution cannot be executed in a pipelined scheme. Therefore, the total latency of the design is approximately 41700 cycles.

Xilinx has a similar CORDIC IP core[43]. Dick *et al.* [44] implemented a complex-valued folded QRD with subsequent back substitution on a Virtex-4 FPGA device using the Xilinx System Generator tool[33]. The folded design contains one diagonal cell (CORDIC-based), one off-diagonal cell (DSP48-based) of the systolic array, and a back-substitution cell with block RAMs and a control unit to schedule the movement of data between blocks. The total cost is approximately 3530 slices, 13 DSP48 blocks and 6 block RAMs. The proposed design to solve a 9×9 system of equations results in approximately 10971 cycles. The area usage of the CORDIC-based QRD is much smaller than that of Dick's 2005[15] 4×4 matrix-inversion core using SGR-based QRD because it uses fixed-point rather than floating-point arithmetic. Furthermore, the SGR operation is based on normal arithmetic operations, while the CORDIC operation is based on multiplication-free and division-free rotation operations[4].

Xilinx also has QR decomposition, QR inverse and QRD-RLS spatial filtering IP cores available in the AccelWare DSP IP Toolkits[45] originally from AccelChip, Inc. AccelWare is a library of floating-point MATLAB model generators that can be synthesized by AccelDSP into efficient fixed-point hardware. These AccelWare DSP cores are used in WiMAX baseband MIMO systems[46] and for beamforming[12] applications. Uribe *et al.* [12] described a 4-element beamformer based on the QRD-RLS algorithm with CORDIC-based Givens rotations. The resources required on the target device (a Xilinx Virtex-4 XC4VSX55 FPGA) are 3076 logic slices and one DSP48 block. The number of slices and DSP48 blocks is further reduced compared to the design in [44] because the design in [12] is mainly based on CORDIC operation while only the diagonal cell is based on CORDIC operation in [44]. The sample throughput rate of the design is quoted at 1.7 MHz. The authors do not state the clock speed of the device they are using, but their chosen device (an XC4VSX55) is available in 400 MHz, 450 MHz and 500 MHz variants, which would give 235,265 and 294 cycles, respectively.

Matousek *et al.* implemented a diagonal cell of the systolic array for QRD using their high-speed logarithmic arithmetic (HSLA) library[47]. In this design, the LNS format is applied, i. e., the number is divided into an integer part, which always has 8 bits, and a fractional part, the size of which depends on the data precision. For 32-bit LNS format, the QRD diagonal cell consumes approximately 3000 slices of Xilinx Virtex devices with 13-cycle latency. For comparison, a floating-point QRD diagonal cell in which data are represented by a 23-bit mantissa and 8-bit exponent is also implemented, requiring 3500 slices with a latency of 84 cycles. The LNS-arithmetic-based design is much faster than the conventional-arithmetic-based design.

Schier *et al.* uses the same HSLA library as that in [47] to implement floating-point operations for Givens rotations[48] and a QRD-based RLS (QRD-RLS) algorithm[49] based on a Xilinx Virtex-E XCV2000E FPGA. Two LNS data formats are implemented: 19-bit and 32-bit. Only one diagonal cell and one off-diagonal cell of the systolic array, not the full array, are implemented. Since addition and subtraction are the most computationally complex modules in an LNS system, they are evaluated using a first-order Taylor-series approximation with look-up tables. Even when using an error-correction mechanism and a range-shift algorithm[50] to minimize the size of the look-up table, the logarithmic addition/subtraction block still requires a large number of slices and memory space. For 19-bit and 32-bit LNS formats, one addition/subtraction block requires approximately 8% and 13% slices and 3% and 70% block RAMs of a single Xilinx Virtex-E XCV2000E FPGA, respectively. Thus, for 19-bit LNS format, these two cells (one diagonal cell and one off-diagonal cell) require approximately 4492 slices and 30 block RAMs. The diagonal cell has a latency of 11 cycles, and the off-diagonal element has a latency of 10 cycles. These two cells can run at 75 MHz and achieve a throughput of approximately 6.8 MHz since both cell types are fully pipelined. By contrast, for 32-bit LNS data format, only one logarithmic addition/subtraction block can fit on the Xilinx XCV2000E FPGA.

Eilert *et al.* [23] implemented a 4 × 4 complex-valued matrix-inversion core in floating-point format for the Xilinx Virtex-4 FPGA based on their BAMI algorithm[22]. The design was implemented using the Xilinx Core Generator[34] to generate all basic units, such as the floating-point real adders, subtractor, real multipliers and real dividers. As the multiplications and divisions are executed using logic gates, not the embedded multipliers or the DSP48 blocks, the BAMI design requires a large chip area. Overall, 7312 slices and 9474 slices are required for 16-bit and 20-bit floating-point data formats, respectively. The 16-bit design and 20-bit design run at 120 MHz and 110 MHz, respectively, both with a latency of 270 cycles.

LaRoche *et al.* [24] synthesized a 4 × 4 complex-valued matrix-inversion unit based on the modified Sherman-Morrison equation (4.9) using Xilinx Synthesis Technology (XST)[51] on a Xilinx Virtex-II XC2V600 FPGA chip. Although this modified Sherman-Morrison equation (4.9) does not contain division operations, its FPGA implementation is remains complicated due to the large number of multiplication operations. The design has four main blocks, a matrix-matrix multiplication block, a matrix-vector multiplication block, a vector-vector multiplication block and a scalar-matrix multiplication block, which consume approximately 3108, 765, 187 and 780 logic slices with 64, 16, 4 and 16 18-bit × 18-bit embedded multipliers, respectively. The total area usage given in [24] is 4446 slices and 101 18-bit × 18-bit embedded multipliers, which is smaller than the sum of the usage of the four main blocks. The author did not provide an explanation for this difference. Moreover, the number of required RAMs is also not given. The latency of the design is approximately 64 cycles.

Table 4-1 Comparison of FPGA-based matrix inversion and linear of equation solvers (MULT = multiplier)

Matrix Size	Technique	Logic Slices	Extras	Cycles
2×2	QRD-SGR[4]	6305	59 MULTs	415
4×4	QRD-SGR[15]	9117	22 DSP48s	933
2×2	QRD-CORDIC[4]	11910	20 MULTs	685
4×4	QRD-CORDIC[4]	16805	44 MULTs	3000
4×4	QRD-CORDIC[12]	3076	1 DSP48s	265
9×9	QRD-CORDIC[36]	1300	1NIOS Processor	41700
9×9	QRD-CORDIC[44]	3530	13 DSP48s	10971
4×4	BAMI[23]	9474	—	270
4×4	Modified Sherman-Morrison[24]	4446	101 MULTs	64
1 diagonal cell	QRD-LNS[47]	3000	—	13
1 diagonal cell	QRD[47]	3500	—	84
1 diagonal cell 1 off-diagonal cell	QRD-LNS[49][48]	4492	—	11 10

Table 4-1 compares some of the FPGA implementations mentioned above. The current approaches to the problem of solving normal equations and matrix inversion demand relatively large computational resources, making notable use of hardware multipliers. Only small-size problems can be efficiently solved in real-time hardware design.

Exercises

1 Explanation of glossary

1-1 QRD

1-2 BAMI

2 Choice question

1-1 The direct methods for solving linear systems include ( ).

A. Gaussian elimination B. Cholesky decomposition

C. LU factorization D. Steepest descent method

2-2 Based on the literature, which technique has been used widely for FPGA based matrix inversion ( ).

A. QRD B. Cholesky decompisition

C. Gauss-Seidel D. LU factorization

3 Short answer question

3-1 Compare direct methods with iterative methods for solving linear systems respectively.

3-2 Talk about what kind of algorithm requiring matrix inversion operation. Is it easy to implement matrix inversion in hardware? Why?

References

[1] HAYKIN S. Adaptive filter theory. 4th ed. New Jersey: Prentice Hall Inc. , 2002.

[2] VAN VEEN B D, Buckley K M. Beamforming: A versatile approach to spatial filtering. IEEE assp magazine, 1988, 5(2):4–24.

[3] ZAKHAROV Y V, TOZER T C. Multiplication – free iterative algorithm for LS Problem. Electronics Letters, 2004, 40(9): 567–569.

[4] MYLLYLA M, HINTIKKA J M, CAVALLARO J R, et al. Complexity analysis of MMSE detector architectures for MIMO OFDM systems. Asilomar Conference on Signals, Systems, and Computers, Pacific Grove, California. 2005:75–81.

[5] KONDOZ A M. Digital speech: coding for low bit rate communication systems. 2nd edition. New York: John Wiley & Sons. , Nov. 2004.

[6] PROAKIS J, MANOLAKIS D. Digital signal processing: principles, algorithms, and applications. 2nd edition. New York: Macmillan, 1992.

[7] SAYED A H. Fundamentals of adaptive filtering. Hoboken, N. J. : Wiley, 2003.

[8] GOLUB G H, VAN LOAN C F. Matrix computations. 3rd edition. Baltimore: The Johns Hopkins University Press, 1996.

[9] LAPACK - linear algebra PACKage. Version 3. 1. 1. 2007[2018-08-25]. http://www. netlib. org/lapack.

[10] WATKINS D J. Fundamentals of matrix computations. Hoboken, N. J.: Wiley, 2004.

[11] Moon T K, Stirling W C. Mathematical methods and algorithms for signal Processing. Upper Saddle River, NJ: Prentice hall, 2000.

[12] URIBE R, CESEAR T. Implementing matrix inversions in fixed-point hardware. Xilinx DSP Magazine, 2005, 1(1): 32 - 25.

[13] White J R. Math Methods -- Section VI: Numerical Solution of Linear Algebraic Equations Lecture Notes, University of Massachusetts Lowell, USA, 2003[2018-08-25]. http://www. profjrwhite. com/math_methods/pdf_files_notes/math_s6. pdf.

[14] LOUKA B, TCHUENTE M. Givens elimination on systolic array. Proceedings of the 2nd International Conference on Supercomputing, St. Malo, France, 1988: 638 - 647.

[15] KARKOOTI M, CAVALLARO J, DICK C. FPGA implementation of matrix inversion using QRD-RLS algorithm. Asilomar Conference on Signals, Systems, and Computers, Pacific Grove, California, 2005: 1625 - 1629.

[16] LIU Z, MCCANNY J V, LIGHTBODY G, et al. Generic SoC QR array processor for adaptive beamforming. IEEE Transactions on Circuits and Systems II: Express Briefs, 2003, 50(4):169 - 175.

[17] EDMAN F, OWALL V. A scalable pipelined complex valued matrix inversion architecture. IEEE International Symposium on Circuits and Systems (ISCAS), Kobe, Japan, 2005(5): 4489 - 4492.

[18] LIGHTBODY G, WOODS R, WALKE R L. Design of a parameterizable silicon intellectual property core for QR-based RLS filtering. IEEE Transactions on Very Large Scale Integration (VLSI) Systems, 2003, 11(4): 659 - 677.

[19] DOHLER R. Squared Givens rotation. IMA Journal of Numerical Analysis, 1991, 11: 1 - 5.

[20] WALKE R L, SMITH R W M, LIGHTBODY G. Architectures for adaptive weight calculation on ASIC and FPGA. Asilomar Conference on Signals, Systems, and Computers, Pacific Grove, California, US., 1999, 2: 1375 - 1380.

[21] WANG X J. Variable precision floating-point divide and square root for efficient FPAG implementation of image and signal processing algorithms. PhD thesis, Northeastern University, USA, Dec. 2007[2018-08-25]. http://www. ece. neu. edu/ groups/rcl/theses/xjwang phd2007. pdf.

[22] EILERT J, DI W, LIU D. Efficient complex matrix inversion for MIMO Software

Definited Radio. IEEE International Symposium on Circuits and Systems (ISCAS), New Orleans, LA, US., May 2007:2610 - 2613.

[23] EILERT J, DI W, LIU D. Implementation of a programmable linear MMSE detector for MIMO - OFDM. IEEE International Conference on Acoustics, Speech, and Signal Processing (ICASSP), Las Vegas, Nevada, US., March 2008: 5396 - 5399.

[24] LAROCHE I, ROY S. An efficient regular matrix inversion circuit architecture for MIMO processing. IEEE International Symposium on Circuits and Systems (ISCAS), Island of Kos, Greece, May 2006.

[25] JIA Y, ANDRIEU C, PIECHOCKI R J, et al. Gaussian approximation based mixture reduction for near optimum detection in MIMO systems. IEEE Communication Letters, Nov. 2005, 9(11): 997 - 999.

[26] BARRETT R, BERRY M, CHAN T F, et al. Templates for the solution of linear systems: building blocks for iterative methods, SIAM, 2nd edition, 1994.

[27] BORAY G K, SRINATH M D. Conjugate gradient techniques for adaptive filtering. IEEE Transactions on Circuits and Systems I: Fundamental Theory and Applications, 1992, 39 (1):1 - 10.

[28] CHANG P S, WILLSON JR A N. Analysis of conjugate gradient algorithms for adaptive filtering. IEEE Transactions on Signal Processing, 2000, 48(2): 409 - 418.

[29] FUKUMOTO M, KANAI T, KUBOTA H, et al. Improvement in the stability of the BCGM - OR algorithm. Electronics and Communications in Japan, Part III: Fundamental Electronic Science, 2000, 83(5): 42 - 52.

[30] DIENE O, BHAYA A. Adaptive filtering algorithms designed using control Liapunov functions. IEEE Signal Processing Letters, 2006, 13(4):224 - 227.

[31] ALBU F, KADLEC J, COLEMAN N, et al. The Gauss-Seidel fast affine projection algorithm. IEEE Workshop on Signal Processing Systems (SiPS), San Diego, CA, US., Oct. 2002: 109 - 114.

[32] ALBU F, BOUCHARD M. The Gauss - Seidel fast affine projection algorithm for multichannel active noise control and sound reproduction systems. International Journal of Adaptive Control and Signal Processing, 2005, 19(2-3):107 - 123.

[33] XILINX Inc. System Generator for DSP. 2007[2018-08-25]. http://www.xilinx.com/ise/optional prod/system generator.htm.

[34] XILINX Inc. Core Generator. 2007 [2018 - 08 - 25]. http://www.xilinx.com/products/design tools/logic design/design entry/coregenerator.htm.

[35] XILINX Inc. Virtex - 4 libraries guide for HDL designs. 2007 [2018 - 08 - 25]. http://toolbox.xilinx.com/docsan/xilinx8/books/docs/v4ldl/v4ldl.pdf.

[36] BOPPANA D, DHANOA K, KEMPA J. FPGA based embedded processing architecture for the QRD-RLS algorithm. Proceeding of the 12th Annual IEEE Symposium on

Field-Programmable Custom Computing Machines (FCCM'04), Apr. 2004:330 - 331.

[37] FITTON M P, PERRY S, JACKSON R. Reconfigurable antenna processing with matrix decomposition using FPGA based application specific integrated processors. Software Defined Radio Technical Conference and Product Exposition (SDR'04), Phoenix, Ariz, US., Nov. 2004.

[38] ALTERA Inc. Application Note 263: CORDIC reference design v1.4, June 2005. 2005[2018-08-25]. http://www.altera.com/literature/an/an263.pdf.

[39] ALTERA Inc. Accelerating WiMAX system design with FPGAs. 2004[2018-08-25]. http://www.altera.com/literature/wp/wp wimax.pdf.

[40] JALALI S. Wireless Channel Equalization in Digital Communication Systems. CGU Theses & Dissertations. 2012[2018-08-27]. http://scholarship.claremont.edu/cgu_etd/42.

[41] CARMICHAEL C, FULLER E, BLAINP, et al. SEU mitigation techniques for Virtex FPGAs in space applications. MAPLD 1999 - Annual Military and Aerospace Applications of Programmable Devices and Technologies Conference, 2nd, Johns Hopkins University, APL, Laurel, MD, USA, Sept. 1999.

[42] ALTERA Inc. VDS Comparison APEX 20KE vs. Virtex-E Devices. August 2000 [2018-08-27]. https://www.intel.com/content/dam/altera-www/global/en_US/pdfs/literature/pib/lvds_comparison1.pdf

[43] XILINX Inc. Xilinx LogiCORE CORDIC v3.0. Xilinx Product Specification DS249, Apr. 2005. [2018-08-25]. http://www.xilinx.com/support/answers/15137.html

[44] DICK C, HARRIS F, PAJICM, et al. Implementing a real-time beamformer on an FPGA platform. XCell Journal, April 2007, 60: 36 - 40.

[45] XILINX Inc. AccelDSP synthesis tool. [2018-08-25]. http://www.xilinx.com/ise/dsp design prod/acceldsp.

[46] HILL T. Researching FPGA implementations of baseband MIMO algorithms using AccelDSP. April 2007[2018-08-25]. http://www.xilinx.com/products/designresources/dspcentral/resource/wimax mimo acceldsp.pdf, April 2007.

[47] MATOUSEK R, TICHY M, POHLZ, et al. Logarithmic number system and floating-point arithmetics on FPGA. Proceedings of the Reconfigurable Computing Is Going Mainstream, 12th International Conference on Field-Programmable Logic and Applications, Montpellier (La Grande-Motte), France, Sept. 2002: 627 - 636,.

[48] SCHIER J, HERMANEK A. Using logarithmic arithmetic for FPGA implementation of the Givens rotations. Proceedings of the Sixth Baiona Workshop on Signal Processing in Communications, Baiona, Spain, Sept. 2003:199 - 204.

[49] SCHIER J, HERMANEK A. Using logarithmic arithmetic to implement the recursive least squares (QR) algorithm in FPGA. 14th International Conference on Field-Programmable Logic and Applications, Leuven, Belgium, August 2004:1149 - 1151.

[50] COLEMAN J N, CHESTER E I, SOFTLEY CI, et al. Arithmetic on the European logarithmic microprocessor. IEEE Transactions on Computer, 2000, 49(7):702 - 715.

[51] XILINX Inc. XST User Guide. UG627(v11.3) September16, 2009[2018-08-27]. https://www.xilinx.com/support/documentation/sw_manuals/xilinx11/xst.pdf

Chapter 5 Multiuser detection

5.1 System models for multiuser detection

5.1.1 Synchronous CDMA systems

In a synchronous CDMA system, the received signal is given by

$$y(t) = \sum_{k=1}^{K} A_k x_k s_k(t) + n(t) \tag{5.1}$$

where

☆ $s_k(t)$ is a signature waveform for the kth user. The signature waveform $s_k(t)$ is represented by

$$s_k(t) = [p(t), p(t - T_c), p(t - (m - 1)T_c)] s_k \tag{5.2}$$

where $p(t)$ is the chip waveform, T_c is the chip period, s_k is a $m \times 1$ vector known as the spreading sequence for user k, and m is the spreading factor[1].

☆ x_k is the input symbol of the kth user.

☆ A_k is the received amplitude of the kth user symbol.

☆ $n(t)$ is the zero-mean additive white Gaussian noise with power spectral density σ^2.

The cross-correlation of the signature waveform is defined as

$$\rho_{ij} = \int_0^{T_s} s_i(t) s_j(t) \mathrm{d}t$$

where T_s is the symbol duration. We define the cross-correlation matrix as: $\boldsymbol{R} = \left\{\rho_{ij}\right\}$, where $\boldsymbol{R}$ is $K \times K$ symmetric non-negative definite matrix.

5.1.2 Asynchronous CDMAsystems

In the case of asynchronous transmission, the received signal is given by[1]

$$y(t) = \sum_{k=1}^{K} \sum_{i=0}^{N} A_k x_k[i] s_k(t - iT_s - \tau_k) + n(t) \tag{5.4}$$

where t is the transmission time, τ_k is the delay in transmission of the kth user, $x_k[i]$ is the ith transmitted symbol of the kth user, and $(N+1)$ is the data block length. We assume that users send a stream of symbols $x_k[0],\cdots,x_k[N]$. If the channel is known, i. e., the receiver knows all the delays τ_k and channel gains A_k, we can consider an asynchronous CDMA system as a special case of the synchronous CDMA when the number of users is $K(N+1)$. The modified spreading sequence is given by

$$\tilde{s}_{k,i}(t) = s_k(t - iT_s - \tau_k) \tag{5.5}$$

The modified channel gain is $\tilde{A}_{k,i} = A_k$, and the modified data symbols are $\tilde{x}_{k,i} = x_k[i]$ where $k = 1,\cdots,K$ and $i = 0,\cdots,N$. The modified signal model can be represented as

$$y(t) = \sum_{k=1}^{K(N+1)} \tilde{A}_k \tilde{x}_k \tilde{s}_k(t) + n(t) \tag{5.6}$$

Therefore, the asynchronous CDMA system can be referred to as a special case of the synchronous CDMA system model in (5.1) when the number of users increases from K to $K(N+1)$.

5.1.3 CDMA systems with flat fading

Slow frequency-flat fading affects the received amplitude without introducing signature waveform distortion. Therefore, the mathematical formula for representing the synchronous CDMA system model (5.1) is also suitable for characterizing the slow-frequency flat fading channels. Fast-frequency flat fading affects the received amplitudes and introduces signature waveform distortion. The modified signature waveform is

$$\tilde{s}_k(t) = A_k(t) s_k(t) \tag{5.7}$$

Hence, the synchronous CDMA model (5.1) is still applicable in fast-frequency flat fading channels:

$$y(t) = \sum_{k=1}^{K} x_k \tilde{s}_k(t) + n(t) \qquad t \in [0, T_s] \tag{5.8}$$

5.1.4 CDMA system with frequency-selective fading

In a frequency-selective fast fading channel, the signature waveform of the kth user undergoes a linear time-variant transformation characterized by an impulse response $r_k(\tau, t)$. The transformed signature waveform in the frequency-selective fading channel $\tilde{s}_k(t)$ can be represented by

$$\tilde{s}_k(t) = \int_{-\infty}^{\infty} r_k(\tau, t) s_k(t - \tau) \mathrm{d}\tau \tag{5.9}$$

Then, the following synchronous CDMA model can be used

$$y(t) = \sum_{k=1}^{K} x_k \tilde{s}_k(t) + n(t) \tag{5.10}$$

Equation (5.10) indicates that the frequency-selective fading channel model can be transformed into a synchronous CDMA system model.

5.2 Multiuser detectors

MAI significantly limits the performance and capacity of transmission systems. Multiuser detection reduces the interference and combats the near-far problems[2]. The optimal, i.e., maximum likelihood (ML), multiuser detector[1] provides detection performance close to that of single-user detection, but the complexity is exponentially proportional to the number of users. The decorrelating detector eliminates the MAI but enhances the noise power[2]. The minimum mean square error (MMSE) detector[1] has better performance than that of the decorrelating detector, but it requires the estimation of amplitudes and matrix inversion. The decision feedback (DF) detector[1] is among the most popular methods because of its simplicity and good performance. Sphere-constrained and box-constrained algorithms allow the solutions to lie within a closed convex set[3], which substantially improves the performance. Semidefinite relaxation (SDR)[4] relaxes the ML problem into a semidefinite problem and provides a BER performance very close to that of the ML detector. The probabilistic data association (PDA) detector treats MAI as Gaussian noise with matched mean and covariance, and offers a detection performance close to that of the ML detector[5, 6]. The sphere-decoding (SD) and branch and bound (BB) detectors achieve optimal performance; however, their worst-case computational complexity is excessive. A comparison of these advanced multiuser detection techniques in[7], in terms of the complexity and detection performance, has shown that the "efficient frontier" of multiuser detectors is primarily composed of the DF detector, PDA detector, and BB detector.

5.2.1 Conventional matched filters

The conventional detector known as the matched filter detector, which correlates the received signal with the desired user's spreading waveform, is presented in figure 5-1.

The output θ_k of the kth matched filter is given by:

$$\theta_k = \int_0^{T_s} y(t) s_k(t) \mathrm{d}t \tag{5.11}$$

where $y(t)$ is the received signal. Therefore, we can write

$$\theta_k = \int_0^{T_s} \{ \sum_{j=1}^{K} A_j x_j s_j(t) + n(t) \} s_k(t) \mathrm{d}t \tag{5.12}$$

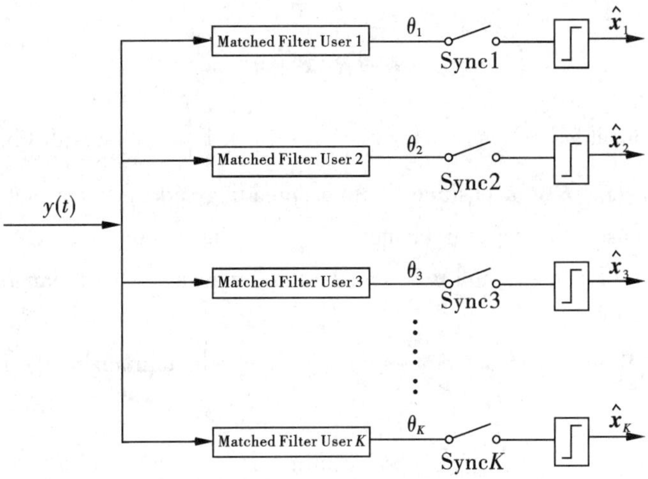

Figure 5-1 Conventional matched filter

By applying equation (5.3) to (5.12), we obtain

$$\theta_k = \sum_{j=1}^{K} A_j x_j \rho_{jk} + n_k \tag{5.13}$$

where

$$n_k = \int_0^{T_s} n(t) s_k(t) \, \mathrm{d}t \tag{5.14}$$

Thus, we have

$$\theta_k = A_k x_k + \sum_{\substack{j=1 \\ j \neq k}}^{K} A_j x_j \rho_{jk} + n_k \tag{5.15}$$

The second term in (5.15) is the MAI. The matched filter treats the MAI as additive white Gaussian noise (AWGN). Nevertheless, the existence of MAI has a substantial effect on the capacity and performance of conventional matched filter detectors.

When the number of interfering users increases, the effect of MAI is significantly enhanced. In addition, users with large amplitudes result in greater MAI effects than those of users with low amplitudes. The signals of closer transmitting users have less amplitude attenuation than that of the signals of transmitting users who are farther away, which is known as the near-far problem[1,8]. The conventional matched filter detector requires no knowledge beyond the spreading sequences. However, as the number of users increases, the matched filter results in poorer detection performance[1].

5.2.2 Maximum likelihood detector

Consider a K-user symbol synchronous CDMA system in an AWGN channel, the output of the matched filter is given by

$$\boldsymbol{\theta} = \boldsymbol{RAx} + \boldsymbol{n} \tag{5.16}$$

where (for BPSK modulation) the vector $\boldsymbol{x} \in \{-1, +1\}^K$ contains the information symbols transmitted by K users, $\boldsymbol{R}$ is a positive definite spreading sequence correlation matrix, $\boldsymbol{A}$ is a diagonal matrix whose kth diagonal element, A_{kk}, is the square root of the received signal energy per bit of the kth user, and $\boldsymbol{n}$ is a real-valued zero-mean Gaussian random vector with covariance matrix $\sigma^2\boldsymbol{R}$。

The optimal ML multiuser detector estimates vector $\boldsymbol{x}$ by minimizing the following quadratic function[1].

$$\hat{\boldsymbol{x}} = \arg \min_{\boldsymbol{x} \in \{-1,+1\}^K} J(\boldsymbol{x}) \tag{5.17}$$

where the quadratic function $J(\boldsymbol{x})$ is represented by

$$J(\boldsymbol{x}) = \| \boldsymbol{\theta} - \boldsymbol{Rx} \|^2 \Rightarrow \boldsymbol{x}^{\mathrm{T}}\boldsymbol{ARAx} - 2\boldsymbol{x}^{\mathrm{T}}\boldsymbol{A\theta} \tag{5.18}$$

The computational complexity of the ML detector is $O(2^K)$ arithmetic operations. Although the ML detector provides the best detection performance, its complexity is exponential in the number of users, which makes hardware implementation challenging.

5.2.3 Decorrelating detector

In model (5.16), the transmitted data can be recovered as

$$\hat{\boldsymbol{x}} = \mathrm{sign}(\boldsymbol{R}^{-1}(\boldsymbol{RAx} + \boldsymbol{n})) = \mathrm{sign}(\boldsymbol{Ax} + \boldsymbol{R}^{-1}\boldsymbol{n}) \tag{5.19}$$

If $\sigma = 0$, then $\hat{\boldsymbol{x}} = \mathrm{sign}(\boldsymbol{x})$. This detector is called the decorrelating detector. The advantage of the decorrelating detector is that it does not require knowledge of the received signal amplitudes. However, when the matrix $\boldsymbol{R}$ is ill-conditioned, the term $\boldsymbol{R}^{-1}\boldsymbol{n}$ in (5.19) results in noise power enhancement, thereby increasing the error probability.

5.2.4 MMSE detector

The MMSE detector takes into account the background noise and uses the knowledge of the received signal power, which leads to a better detection performance than that of the decorrelating detector for low SNRs. The MMSE detector minimizes the mean squared error between the actual symbol and the decision output of the detector[9]. The inverse of matrix $\boldsymbol{R}$ in the decorrelating detection is replaced by $[\boldsymbol{R} + \sigma^2\boldsymbol{A}^{-2}]^{-1}$.

The MMSE detector has a trade-off between MAI elimination and noise enhancement. This detector achieves performance similar to that of the conventional MF when the noise variance approaches infinity. When the SNR goes to infinity, the MMSE detector converges to the decorrelating detector[1, 10]. The MMSE detector is resistant to the near-far problem.

5.2.5 Decision feedback detector (DF detector)

DF detectors have been applied to successive interference cancellation[11-13], parallel interference cancellation[14-16] and multistage or iterative DF detectors[15-16]. The DF detector with successive interference cancellation (S-DF) is optimal in the sense that it achieves the sum capacity of the synchronous AWGN channel[12]. The S-DF scheme can alleviate the effects of error propagation despite often resulting in non-uniform performance over the users. In particular, user ordering plays an important role in the performance of S-DF detectors. Studies on decorrelator DF detectors with optimal user ordering have been reported in[13] for imperfect feedback and in[17] for perfect feedback. The problem with the optimal ordering algorithms in[13] and[17] is that they represent a very high computational burden for practical receiver design. By contrast, the DF receiver with parallel interference cancellation (P-DF)[14-16] satisfies the uplink requirements, i. e., cancellation of intracell interference and suppression of the remaining other-cell interference, and generally provides uniform performance over the user population, even though it is more sensitive to error propagation. The multistage and iterative DF schemes presented in [15], [16] are based on the combination of S-DF and P-DF schemes in multiple stages to refine the symbol estimates, resulting in improved performance compared to that of the conventional S-DF and P-DF and mitigation of error propagation[18].

5.2.6 Semidefinite relaxation detector

Relaxation is an effective approximation technique for certain difficult optimization problems. Relaxing some constraints simplifies the solution to the problem. The semidefinite relaxation algorithm reduces the computational complexity without a performance loss. The semidefinite relaxation algorithm for solving the Boolean quadratic-programming (QP) problem[19-21] is described first,

$$\arg\min_{\boldsymbol{x}}(\boldsymbol{x}^{\mathrm{T}}\boldsymbol{Q}\boldsymbol{x}) \tag{5.20}$$

where $\boldsymbol{Q}$ is any symmetric matrix. Since $\boldsymbol{x}^{\mathrm{T}}\boldsymbol{Q}\boldsymbol{x} = \mathrm{Trace}(\boldsymbol{x}\boldsymbol{x}^{\mathrm{T}}\boldsymbol{Q})$, problem (5.20) can be restated as

$$\begin{aligned} &\arg\min_{\boldsymbol{X}}\mathrm{Trace}(\boldsymbol{Q}\boldsymbol{X}) \\ &\text{s. t. } \operatorname{diag}(\boldsymbol{X}) = \boldsymbol{e} \\ &\boldsymbol{X} = \boldsymbol{x}\boldsymbol{x}^{\mathrm{T}} \end{aligned} \tag{5.21}$$

where $\boldsymbol{e}$ is a vector composed of all ones.

The constraint $\boldsymbol{X} = \boldsymbol{x}\boldsymbol{x}^{\mathrm{T}}$ implies that $\boldsymbol{X}$ is symmetric, positive semidefinite and of rank 1. Due to the constraint $\boldsymbol{X} = \boldsymbol{x}\boldsymbol{x}^{\mathrm{T}}$, problem (5.21) is a non-convex optimization problem. When the rank-1 constraint is removed from (5.21), we obtain the following problem

$$\arg \min_{\boldsymbol{X}} \mathrm{Trace}(\boldsymbol{QX})$$
$$\text{s.t. } \boldsymbol{X} \succeq 0 \tag{5.22}$$
$$X_{jj} = 1, j = 1, \cdots, K$$

where $\boldsymbol{X} \succeq 0$ means that $\boldsymbol{X}$ is symmetric and positive semidefinite. Problem (5.22) is a relaxation of problem (5.21) because the feasible set in (5.21) is a subset of that in (5.22). The problem in (5.22) is considered to be the semidefinite relaxation of (5.21). The problem of (5.22) can be solved on the order of $O(K^{3.5})$ operations.

To apply the semidefinite relaxation algorithm to the ML detection problem, theoriginal ML detection problem has to be rewritten in the same form as (5.20). Define a scalar $c \in \{-1, +1\}$. Since $c\boldsymbol{x} \in \{-1, +1\}^K$ for any $\boldsymbol{x} \in \{-1, +1\}^K$, (5.17) can be rewritten as

$$\begin{aligned} \max_{\boldsymbol{x} \in \{-1,+1\}^K} J(\boldsymbol{x}) &\equiv \max_{\substack{\boldsymbol{x} \in \{-1,+1\}^K \\ c \in \{-1,+1\}}} J(c\boldsymbol{x}) \\ &= \max_{\substack{\boldsymbol{x} \in \{-1,+1\}^K \\ c \in \{-1,+1\}}} 2c\,\boldsymbol{x}^{\mathrm{T}}\boldsymbol{A}\boldsymbol{y} - \boldsymbol{x}^{\mathrm{T}}\boldsymbol{ARA}\boldsymbol{x} \\ &= \max_{\substack{\boldsymbol{x} \in \{-1,+1\}^K \\ c \in \{-1,+1\}}} \begin{bmatrix} \boldsymbol{x}^{\mathrm{T}} & c \end{bmatrix} \begin{bmatrix} -\boldsymbol{ARA} & \boldsymbol{Ay} \\ (\boldsymbol{Ay})^{\mathrm{T}} & 0 \end{bmatrix} \begin{bmatrix} \boldsymbol{X} \\ c \end{bmatrix} \end{aligned} \tag{5.23}$$

This equation is equivalent to the Boolean QP problem in (5.20) and

$$\boldsymbol{Q} = \begin{bmatrix} -\boldsymbol{ARA} & \boldsymbol{Ay} \\ (\boldsymbol{Ay})^{\mathrm{T}} & 0 \end{bmatrix}$$

The semidefinite relaxation detector provides a BER performance close to that of the ML detector, even when the cross-correlations between users are strong or the near-far effect is significant. However, the semidefinite relaxation detector is very complex for large systems[7,20].

5.2.7 Constrained multiuser detectors

The ML detector for BPSK modulation finds a solution constrained to $\boldsymbol{x} \in \{-1, +1\}^K$, where $\{-1, +1\}^K$ denotes the set of all binary symbols and each symbol is either +1 or −1. However, the ML detector has been proved to be too complex for practical use[2]. A simple limitation of constraining the symbol estimate vector to within a closed convex set often reduces the complexity. Constraining the symbol estimate to lie within a hypercube results in a box-constrained quadratic problem (e.g., $K = 2$), as shown in figure 5-2(a)[22]. When applying the box constraint to solve the ML problem in a CDMA system with BPSK, the problem is reformulated as

$$\hat{\boldsymbol{x}} = \arg \min_{\boldsymbol{x} \in [-1,+1]^K} J(\boldsymbol{x}) \tag{5.24}$$

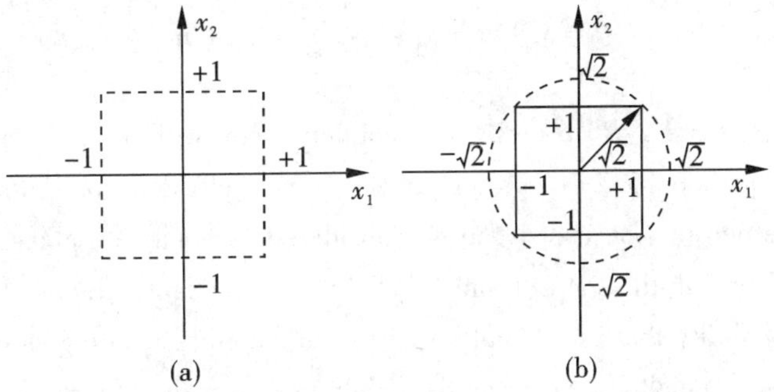

Figure 5-2 Projection onto a box region(a) and projection onto a sphere region(b)

We define an orthogonal projection operation P_B on a hypercube closed convex set $\beta(\beta=[-1,+1])$, which is represented as

$$P_B(\hat{\boldsymbol{x}}) = \arg\min_{\boldsymbol{x}\in\beta} \| \boldsymbol{x} - \hat{\boldsymbol{x}} \| \tag{5.25}$$

where $P_B(x_k)$ is represented as

$$\begin{cases} x_k & \text{if} \quad -1 < x_k < 1 \\ -1 & \text{if} \quad x_k \leqslant -1 \\ +1 & \text{if} \quad x_k \geqslant +1 \end{cases} \tag{5.26}$$

The sphere-constrained ML problem[23] is described in figure 5-2(b):

$$\hat{\boldsymbol{x}} = \arg\min_{\hat{\boldsymbol{x}}\in S} J(\boldsymbol{x}) \tag{5.27}$$

where $S = \{\boldsymbol{x} \in \mathcal{R}^K: \| \boldsymbol{x} \|^2 \leqslant K\}$. We assume $P_S(x_k)$ is the kth element of the orthogonal projection onto a sphere, which is represented as

$$\begin{cases} \alpha x_k & \text{if} \quad \| \boldsymbol{x} \|^2 > K \\ x_k & \text{if} \quad \| \boldsymbol{x} \|^2 \leqslant K \end{cases} \tag{5.28}$$

where $\alpha = \sqrt{K}/ \| \boldsymbol{x} \|$ and $0 < \alpha \leqslant 1$. A multiuser detector with a sphere constraint provides detection performance similar to that of the MMSE detector[24]. Furthermore, figure 5-2(b) shows that the sphere region contains the box region $[-1, +1]$, which means the detector with the box constraint is more restricted, thus resulting in better performance than that of the sphere-constrained detector[7,24,25].

5.2.8 Probabilistic data association detector

The result of multiplying $\boldsymbol{A}^{-1}\boldsymbol{R}^{-1}$ by both sides of (5.16), the equation can be represented as[6]

$$\bar{\boldsymbol{\theta}} = \boldsymbol{x} + \bar{\boldsymbol{n}} = x_i \boldsymbol{e}_i + \sum_{j \neq i} x_j \boldsymbol{e}_j + \bar{\boldsymbol{n}} \tag{5.29}$$

where $\bar{\boldsymbol{\theta}} = \boldsymbol{A}^{-1}\boldsymbol{R}^{-1}\boldsymbol{\theta}, \bar{\boldsymbol{n}} = \boldsymbol{A}^{-1}\boldsymbol{R}^{-1}\boldsymbol{n}$ and $\boldsymbol{e}_i$ is a column vector with a one in the ith position and zeros elsewhere. Equation (5.29) is a normalized version of the decorrelator output before the hard decision. The decision on user i can be considered as binary variables +1 or −1 with the currently estimated probabilities $P_x(i)$ and $1 - P_x(i)$, respectively, i. e., $P_x(i)$ is the current estimate of the probability that $x_i = 1$ and $1 - P_x(i)$ is the corresponding estimate for $x_i = -1$. For an arbitrary user signal x_i, we treat the other user signals $x_j (j \neq i)$ as binary random variables and treat $\sum_{j \neq i} x_j \boldsymbol{e}_j + \bar{\boldsymbol{n}}$ as the effective noise. Based on the decorrelated model, the basic multistage PDA algorithm is described as follows[6].

1. Sort users according to the user ordering principle for the DF detector in[13].
2. For all users, initialize $P_x(i) = 0.5$ and stage counter $k = 1$.
3. Initialize the user counter $i = 1$.
4. According to the current $P_x(j) (j \neq i)$ for user i, update $P_x(i)$ by

$$P_x(i) = P\left\{ x_i = 1 \,\middle|\, \bar{\boldsymbol{\theta}}, \left\{ P_x(j) \right\}_{j \neq i} \right\} \tag{5.30}$$

5. If $i < K$, let $i = i + 1$ and go to step 4.
6. If $\forall i, P_x(i)$ has converged, go to step 7. Otherwise, let $k = k + 1$ and return to step 3.
7. make a decision on user signal i, i. e., x_i, via

$$\begin{cases} 1 & P_x(i) \geq 0.5 \\ -1 & P_x(i) < 0.5 \end{cases} \tag{5.31}$$

The computational cost of obtaining (5.31) is exponential in the number of users. We define

$$\breve{\boldsymbol{n}}_i = \sum_{j \neq i} x_j \boldsymbol{e}_j + \bar{\boldsymbol{n}} \tag{5.32}$$

We denote the mean and covariance matrix of $\breve{\boldsymbol{n}}_i$ as

$$\begin{aligned} E(\breve{\boldsymbol{n}}_i) &= \sum_{j \neq i} \boldsymbol{e}_j (2P_x(j) - 1) \\ \mathrm{Cov}(\breve{\boldsymbol{n}}_i) &= \sum_{j \neq i} \left[4P_x(j)(1 - P_h(j))\, \boldsymbol{e}_j \boldsymbol{e}_j^{\mathrm{T}} \right] + \sigma^2 \boldsymbol{R}^{-1} \end{aligned} \tag{5.33}$$

Correspondingly, we define $\boldsymbol{\Phi}_i = E(\breve{\boldsymbol{n}}_i)$ and $\boldsymbol{\Psi}_i = \mathrm{Cov}(\breve{\boldsymbol{n}}_i)$. The updated probability $P_x(i)$ is given by

$$\frac{P_x(i)}{1 - P_x(i)} = \exp\left\{ -2 \boldsymbol{\Phi}_i^{\mathrm{T}} \boldsymbol{\Psi}_i^{-1} \boldsymbol{e}_i \right\} \tag{5.34}$$

The algorithm continues until all the probabilities $\{P_{x_i}\}$ have converged. The detection performance approaches the single-user performance over high SNRs. More details are shown in [6]. The overall complexity of the PDA detector is $O(K^3)$ [26,27].

5.2.9 Dichotomous coordinate descent (DCD) algorithm

In many communications systems, detection is based on the solution of system $\boldsymbol{Rx} = \boldsymbol{\theta}$, where $\boldsymbol{R}$ is a $K \times K$ symmetric positive definite matrix and both $\boldsymbol{x}$ and $\boldsymbol{\theta}$ are $K \times 1$ vectors. The matrix $\boldsymbol{R}$ and vector $\boldsymbol{\theta}$ are known, whereas solution $\boldsymbol{x}$ is unknown.

Table 5-1 Dichotomous Coordinate Descent Algorithm

Initialization: $\boldsymbol{x} = \bar{\boldsymbol{x}}$, $\boldsymbol{r} = \boldsymbol{\theta} - \boldsymbol{R}\bar{\boldsymbol{x}}$, $d = H$, $p = 0$
for $m = 1:M_b$ 1. $d = d/2$ 2. **Flag** = 0 ⋯pass 3. **for** $j = 1:K$ ⋯iteration 4. **if** $\vert r(j) \vert > (d/2)R(j,j)$ **then** 5. $x(j) = x(j) + \text{sign}(r(j))d$ 6. $\boldsymbol{r} = \boldsymbol{r} - \text{sign}(r(j)) \cdot d \cdot \boldsymbol{R}(:,j)$ 7. **Flag** = 1, $p = p + 1$ 8. **if** $p > N_u$ **end algorithm** 9. **end** j**-loop** 10. **if Flag** = 1, goto 2 **end** m**-loop**

The DCD algorithm[28] is designed to provide a simple solution for vector $\boldsymbol{x}$ without explicit multiplication and division. The accuracy of solution vector $\boldsymbol{x}$ depends on the number of bits (M_b) used to represent the elements of vector $\boldsymbol{x}$ within an amplitude range $[-H, H]$. The first set of iterations in the algorithm determines the most significant bit ($m = 1$) for all elements of $\boldsymbol{x}$ using a step size parameter d. The subsequent sets of iterations determine the less significant bits up to a suitable number of bits (a maximum of M_b). The initial residual vector is given by $\boldsymbol{r} = \boldsymbol{\theta} - \boldsymbol{R}\bar{\boldsymbol{x}}$, where $\bar{\boldsymbol{x}}$ is the initialization of $\boldsymbol{x}$. Table 5-1 describes the DCD algorithm. We denote the elements of vectors $\boldsymbol{x}$ and $\boldsymbol{r}$ as $x(j)$ and $r(j)$, respectively. When the vector $\bar{\boldsymbol{x}}$ is set to zero, $\boldsymbol{r}$ is equal to $\boldsymbol{\theta}$. The step size is set to H, and the successful iteration counter p is set to 0. The step size d is reduced by a power of two in step 1; therefore, no explicit multiplication or division is required because the multiplication and division can be replaced by simple bit shifts. If an element of the solution vector is updated in an iteration, such an iteration is labeled "successful". For every step size update, the algorithm repeats successful iterations until all elements of the residual vector $\boldsymbol{r}$ are sufficiently small that the condition in step 4 is not met for all j. The computational load of the algorithm depends mainly on these successful update iterations p and the number of bits M_b. A limit for the number of successful iterations N_u can be predefined and used as a stopping condition. If there is no such limit, or the limit is sufficiently high, the accuracy of the solution is 2^{-M_b+1}. The complexity of the DCD algorithm for a particular system of equations depends on many factors. However, for given N_u and M_b, the worst-case scenario complexity is presented as $K(2N_u + M_b)$ shift-accumulation (SAC) operations[29].

5.2.10 Lattice detection: branch and bound detector

The universal lattice decoding problem dates back to at least the early 1990s[30]. The principle of universal lattice decoding can trace its roots to the theory and algorithms developed to solve the shortest/closest lattice vector problem for integer programming problems. The closest (lattice) vector problem (CVP) (also called the nearest lattice point problem) is a class of nearest-neighbor searches or closest-point queries, in which the solution set to be searched consists of all the points in a lattice. Very efficient algorithms for solving the CVP[31] have been derived for the root lattices, which are generated by the root system of certain Lie algebras. These algorithms are important for implementing low-complexity lattice quantizers and coding schemes for Gaussian channels. From a lattice point of view, ML decoding corresponds to solving the closest vector problem in a lattice. However, optimal multiuser detection is NP-hard, but some suboptimal algorithms that are solvable with polynomial complexity exist. BB is a divide-and-conquer structure for the hard combinatorial optimization problem[32]. The main idea is to separate the solution set of a discrete optimization problem into successively smaller subsets (branches), bound the cost function value over each subset and use the bounds to remove some subsets. The process stops when the entire solution set has been searched. The best solution of the BB algorithm is a global optimum since it effectively searches the whole solution space. This method is most efficient when it is possible to remove many subsets as early as possible during the branching procedure without calculating them. Luo, *et al.* [33] proposed a detector based on depth-first BB[34] and showed that the sphere decoder is a type of depth-first BB[35]. The BB can be seen as a tree search algorithm, where each subset is represented by a node in a tree, and the root of the tree represents the whole solution space. Each node is related to a cost that is a lower bound of the global optimum. Accordingly, if a node cost exceeds the current best solution, the node can be pruned, i. e., the children of the node can be removed without loss. The algorithm maintains a list (queue) of nodes to be processed. When a node is retained, its children branches are made and their costs are evaluated. Children nodes whose cost is less than the current best solution are added to the list. Unfortunately, this algorithm may have to save the whole tree in a worst-case scenario. The memory requirements are exponential in terms of the number of levels in the search tree[36].

Table 5-2 Comparison of multiuser detection algorithms

Name	Complexity	Benefits	Shortcomings
Maximum Likelihood	$O(2^K)$	Optimal detection performance	Exponential computational complexity
Matched Filter	$O(K)$	Simplicity	Poor performance
Decorrelating Detector	$O(K^3)$	No requirement for knowledge of the power of the interference	Requires matrix inversion
MMSE	$O(K^3)$	Better performance than that of the decorrelating detector at low SNRs	Requires matrix inversion, estimation of user's amplitude and noise variance
Decorrelating Decision Feedback Detector	$O(K^2)$ (no Cholesky decomposition)	Better performance than that of the linear detectors	Performance strongly relies on the detection order. Requires Cholesky decomposition and matrix inversion
Semidefinite Relaxation	$O(K^{3.6})$	Close to ML detector performance	Complexity is too high for large systems
Sphere-constrained Detector	$O(K^3)$	Close to the MMSE detector	Not optimal performance
Box-constrained Detector	$O(K^3)$	Similar to the soft interference cancellation; Better performance than that of the sphere-constrained detector	Not optimal performance

Table 5-2

Name	Complexity	Benefits	Shortcomings
Probabilistic Data Association	$O(K^3)$	Close to the single-user performance	Requires matrix inversion
Branch and Bound	The worst-case complexity is exponential in K	Low average complexity	Worst-case complexity is too high at low SNRs
Dichotomous Coordinate Descent	$K(2N_u + M_b)$	Multiplication and division free; Efficient for FPGA implementation	Not optimal performance

As a summary, table 5-2 compares some multiuser detection algorithms. The benefits, complexity and shortcomings of these detectors are emphasized in this table. The optimal ML detector unfortunately has exponential complexity $O(2^K)$. The traditional matched filter detector is taken straight from single-user design with a low complexity $O(K)$. However, since it does not take into account any other users in the system, it cannot provide good performance. The decorrelating detector essentially applies the inverse of the correlation matrix of user spreading sequences to the output of the conventional detector. The complexity of the decorrelating detector is $O(K^3)$. This detector does not require the power of each user estimation or control; however, the noise is increased by $\boldsymbol{R}^{-1}\boldsymbol{n}$. The MMSE detector minimizes the squared error in the presence of channel noise and has better performance than that of the decorrelator at low SNRs. However, it requires matrix inversion, which is complicated in FPGA implementations. The DF detector is among the most popular methods in multiuser detection because of its simplicity and outstanding performance compared with those of the linear detectors. However, its performance mainly depends on the detection order. The zero-forcing DF detector requires Cholesky decomposition and matrix inversion, which are difficult to implement. The semidefinite relaxation detector is a complexity-constrained alternative for the exact ML detector, but the complexity remains high in large systems. The box-constrained detector corresponds to nonlinear successive and parallel interference cancellation structures. The sphere-constrained ML detector is closely related to the MMSE detector. The PDA detector approaches the single-user detector performance, but it requires matrix inversion to obtain the covariance of the noise. The DCD detector is multiplication and division free, which is efficient for hardware implementation, but its performance is not as good as that of the PDA detector. The BB detector has low average complexity at high SNRs, but its worst-case computational complexity is identical to that of the optimal multiuser detector, i. e. , it grows exponentially with increasing K.

Exercises

1 Explanation of glossary

1-1 Near-far problem

1-2 SAC

1-3 FPGA

2 Choice question

2-1 ( ) affects the received amplitudes and introduces signature waveform distortion.

A. Fast frequency-flat fading B. Slow frequency-flat fading

C. Frequency-selective fast fading D. Frequency-selective slow fading

2 - 2 ( ) reduces the average complexity and providing the near ML performance .

A. Maximum Likelihood detector B. Branch and bound

C. MMSE detector D. Decorrelating detector

3 Short answer question

3-1 Analyze the relationship between flat fading and frequency selective fading.

3-2 Give a description of Lattice detection and find one Lattice detection alogorithm for example to explain how they work.

Reference

[1] VERDU S. Multiuser Detection. 2nd ed. New York, NY, USA: Cambridge University Press, 1998.

[2] VERDU S. Minimum probability of error for asynchronous Gaussian multiple access channels. IEEE Trans. Inform. Theory, Jan. 1986, 32: 85 - 96.

[3] TAN P H, RASMUSSEN L K, LIM T J. Constrained maximum-likelihood detection in CDMA. IEEE Trans. Commun. , Jan. 2001, 49(1): 142 - 153.

[4] MA W K, CHING T N, DING Z. Semidefinite relaxation based multiuser detection for M-ary PSK multiuser systems. IEEE Trans. Signal Process. , Oct. 2004, 52(10) :2862 - 2872.

[5] WANG X M, LU W S, ANTONIOU A. Multiuser detectors for synchronous DSCDMA systems based on a recursive p-norm convex relaxation approach. IEEE Transactions on Circuits and Systems I: Regular Papers, May 2005, 52(5):1021 - 1031.

[6] LUO J, PATTIPATI K R, WILLETT PK, et al. Near-optimal multiuser detection in synchronous CDMA using probabilistic data association. IEEE Comm. Letters, Sept. 2001, 5 (9):361 - 363.

[7] HASEGAWA F, LUO J, PATTIPATIK, et al. Speed and accuracy comparison of techniques for multiuser detection in synchronous CDMA. IEEE Trans. Commun. , Apr. 2004, 52(4): 540 - 545.

[8] MOSHAVI S. Multi - user detection for DS - CDMA communications. IEEE Communications Magazine. , Oct 1996, 34(10):124 - 136.

[9] GHARSALLAH R, BOUALLEGUE R. Combined ML - MMSE receiver of an STBCCDMA system for PSK/QAM modulation. International Journal of Computer Science and Network Security, Jan. 2007, 7(1): 254 - 258.

[10] CAMPOS - DELGADO D U, MARTINEZ - LOPEZ F J, LUNA - RIVERA J M. Analysis and performance evaluation of linear multiuser detectors in the downlink of DS-CDMA systems applying spectral decomposition. Circuits Systems Signal, Processing, 2007, 26(5):

689 - 713.

[11] DUEL - HALLEN A. A family of multiuser decision - feedback detectors for asynchronous CDMA channels. IEEE Trans. Commun. , Feb. -Apr. 1995, 43: 421 - 434.

[12] VARANASI M K, GUESS T. Optimum decision feedback multiuser equalization with successive decoding achieves the total capacity of the Gaussian multiple access channel. In Proc. 31st Asilomar Conf. Signals, Systems and Computers, Monterey, 2: 1405 - 1409.

[13] VARANASI M K. Decision feedback multiuser detection: a systematic approach. IEEE Trans. Inform. Theory, Jan. 1999, 45(1):219 - 240.

[14] WOODWARD G, RATASUK R, HONIG ML, et al. Multistage multiuser decision-feedback detection for DS-CDMA. In Proc. IEEE ICC, June 1999, 1: 68 - 72.

[15] WOODWARD G, RATASUK R, HONIG ML, et al. Minimum mean-squared error multiuser decision-feedback detectors for DS-CDMA. IEEE Trans. Commun. , Dec 2002, 50 (12): 2104 - 2112.

[16] HONIG M, WOODWARD G, SUN Y. Adaptive iterative multiuser decision feedback detection. IEEE Trans. WirelessCommun. , Mar. 2004, 3(2): 477 - 485.

[17] LUO J, PATTIPATI K R, WILLET PK, et al. Optimal user ordering and time labeling for ideal decision feedback detection in asynchronous CDMA. IEEE Trans. Commun. , Nov. 2003, 51(11): 1754 - 1757.

[18] De LAMARE R C, SAMPAIO-NETO R. Minimum mean-squared error iterative successive parallel arbitrated decision feedback detectors for DS - CDMA systems. IEEE Transactions on Communications, May 2008, 56(5): 778 - 789.

[19] MA W K, DAVIDSON T N, WONG KM, et al. Quasi-maximum- likelihood multiuser detection using semi-definite relaxation with application to synchronous CDMA. IEEE Trans. Signal Processing, Apr. 2002, 50(4): 912 - 922.

[20] JALDEN J, OTTERSTEN B. The diversity order of thesemidefinite relaxation detector. IEEE Transaction on Information Theory, April 2008, 54(4): 1046 - 1422.

[21] MA W K, DAVIDSON T N, WONG K M, et al. Efficient quasi-maximum-likelihood multiuser detection by semi-definite relaxation. IEEE International Conference on Communications, ICC 2001, 2001: 11 - 14.

[22] PARDALOS P M, RESENDE M G C. Interior point methods for global optimization. In Interior Point Method of Mathematical Programming, T. Terlaky, Ed. Norwell, MA: Kluwer, 1996, 5: 467 - 500.

[23] TAN P H, RASMUSSEN L K, LIM T J. Sphere-constrained maximum-likelihood detection in CDMA. In Proc. 2000. Int. Zurich Sem. Broadband Communications, Zurich, Switzerland, Feb. 2000: 55 - 62.

[24] TAN B P, RASMUSSEN L K, LIM TJ. Constrained maximum-likelihood detection in CDMA, IEEE Trans. Commun, Jan. 2001, 49(1): 142 - 153.

[25] TAN P H, RASMUSSEN L K, LIM T J. Box-constrained maximum-likelihood

detection in CDMA. In Proc. IEEE 51st Vehicular Technology Conf. Spring- 2000, Tokyo, Japan, May 2000, 1: 517 - 521.

[26] LUO J. Improved multiuser detection in code - division multiple access communications, PhD Thesis in University of Connecticut, 2002.

[27] TAN P H, RASMUSSEN L K, LUO J. Iterative multiuser decoding based on probabilistic data association. In Proc. IEEE InternationalCommun. Lett. , June 2003: 301 - 301.

[28] ZAKHAROV Y V, TOZER T C. Multiplication - free iterative algorithm for LS problem. Electronics Letters, Apr. 2004, 40(9): 567 - 569.

[29] ZAKHAROV Y V, ALBU F. Coordinate Descent Iterations in Fast Affine Projection Algorithm. IEEE Signal ProcessingLett. , 2005, 12:353-356.

[30] VITERBO E, BOUTROS J. A universal lattice code decoder for fading channels. IEEE Transactions on Information Theory, Jul. 1999, 45(5): 1639 - 1642.

[31] CONWAY J H, SLOANE N J A. Fast quantizing and decoding algorithms for latticequantizers and codes. IEEE Trans. Inform. Theory, Mar. 1982, 28(2): 227 - 232.

[32] PAPADIMITRIOU C H, STIEGLITZ K. Combinatorial Optimization: Algorithms and Complexity. Englewood Cliffs, NJ: Prentice Hall, 1982.

[33] LUO J, PATTIPATI K, WILLETTP, et al. Branch-and-bound-based fast optimal algorithm for multiuser detection in synchronous CDMA. In Proceedings of the IEEE International Conference on Communications, May 2003, 5: 3336 - 3340.

[34] BERTSEKAS D. Network Optimization: Continuous and Discrete Models. Belmont, MA. : Athena Scientific , 1998.

[35] LUO J, PATTIPATI K P, WILLETP, et al. Fast optimal and suboptimal any-time algorithms for CDMA multiuser detection based on branch and bound. IEEE Trans. Commun. , 2004, 52(4): 632 - 641.

[36] FOULADI S, TELLAMBURA C. Smart maximum - likelihood CDMA multiuser detection. 2005 IEEE Pacific Rim Conference on Communications, Computers and signal Processing, PACRIM. , 2005: 522 - 525.

Chapter 6 MIMO systems

6.1 Basic concepts

A conventional wireless system is a single-input single-output (SISO) antenna system that is limited by the channel capacity. Regardless of the modulation scenario, there is still a physical restriction of a single wireless channel. To increase the channel capacity, more BSs, transmission power, or bandwidth are required. Thus, an antenna array or several independent antennas can be applied at the receiver to increase the receiving diversity while the transmitter is equipped with a single antenna. This is called a single-input multiple-output system. An antenna array or several independent antennas can also be included in the transmitter, and the receiver is equipped with a single antenna to decrease the complexity of the receiver. This type of system is called a multiple-input single-output system[1].

Furthermore, when both the transmitter and the receiver have an antenna array or several independent antennas, the system is called a multiple-input multiple-output (MIMO) system. MIMO techniques not only offer high spectral efficiency or high data rates, which indicate more bits per second per hertz of bandwidth, but also have the merits of reliability and diversity. Although MIMO systems are complex, they are an integral component of modern wireless communication standards, such as 4G, Worldwide Interoperability for Microwave Access (WiMAX), and IEEE 802.11n. Figure 6-1 shows the four types of antenna system models.

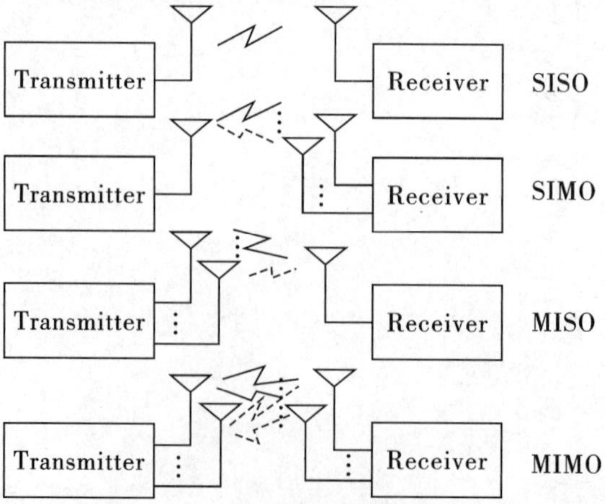

Figure 6-1 Antenna system models

6.2 MIMO systems model

Using multiple antennas at both ends of a communication system can dramatically enhance the system capacity[1]. A schematic diagram of a MIMO system is given in figure 6-2.

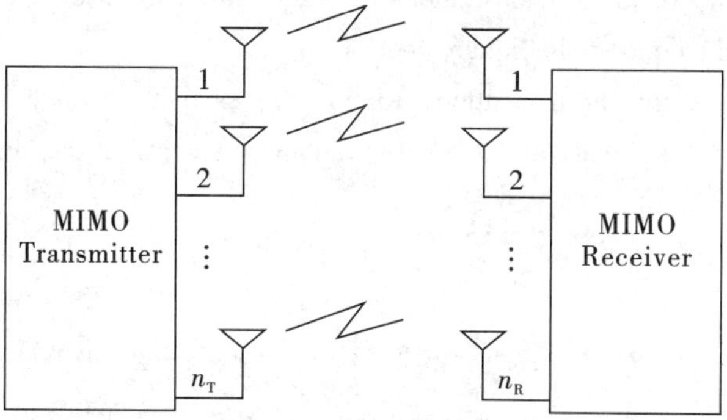

Figure 6-2 Schematic diagram of a MIMO system

In spatial multiplexing (SM), a series of data are first demultiplexed into substreams; then, the information substreams are sent with different preprocessing (coding, modulation, delay, etc.) from different antennas. The data substreams are transmitted through each transmit antenna simultaneously[2]. However, a suitable algorithm is needed at the receiver to separate the signals received on multiple antennas and to recover the original transmitted data stream to maximize the transmission data rate and capacity.

In space-time coding (STC), a number of code symbols equal to the number of transmit antennas is generated and transmitted simultaneously, one symbol from each antenna. These symbols are generated by the space-time encoder[3] or via carefully designed orthogonal transmission of symbols in space and time, such that by using the appropriate signal processing and decoding procedure at the receiver, the diversity gain and/or coding gain is maximized.

We consider a MIMO system with arrays of n_T transmit antennas and n_R receive antennas. $\boldsymbol{x} = [x_1, x_2, \cdots, x_{n_T}]^T$ represent the data on each transmit antenna during one transmission period, and $\boldsymbol{r} = [r_1, r_2, \cdots, r_{n_R}]^T$ represent the data at the receiver. The received signal can be written in the form:

$$\boldsymbol{r} = \boldsymbol{H}\boldsymbol{x} + \boldsymbol{n} \tag{6.1}$$

where $\boldsymbol{H}$ represents the channel matrix and can be expressed as

$$H = \begin{bmatrix} h_{11} & h_{12} & \cdots & h_{1n_T} \\ h_{21} & h_{22} & \cdots & h_{2n_T} \\ \cdots & \cdots & & \cdots \\ h_{n_R1} & h_{n_R2} & \cdots & h_{n_Rn_T} \end{bmatrix} \tag{6.2}$$

where $h_{ij}(i = 1,2,\cdots,n_R; j = 1,2,\cdots,n_T)$ denotes the fading factor of the channel from transmit antenna j to receive antenna i. Assuming we use the independent Rayleigh channel model, h_{ij} is a complex random Gaussian variable of the form:

$$h_{ij} = x + y\iota \tag{6.3}$$

where x and y are independent real random Gaussian numbers with a mean of zero and a variance of 0.5, and ι denotes the square root of -1.

In general, we assume normalization, which ensures that the total receive power is the same as the total transmit power, averaged over random instances of the channel matrix, that is

$$\sum_{i=1}^{n_R}\sum_{j=1}^{n_T}\left|h_{ij}\right|^2 = n_T \tag{6.4}$$

In equation (6.1), $\boldsymbol{n} = \left[n_1, n_2, \cdots, n_{n_R}\right]^T$ represents the channel noise, which is the AWGN noise with noise power σ^2 per dimension.

6.3 Diversity and BER performance

Traditionally, multiple antennas have been used to increase diversity to combat channel fading. For example, we may employ a single transmit antenna and n receive antennas. By sending the same information through $n_R n_T$ different paths, multiple independently faded replicas of the transmitted signal can be obtained at the receiver. Therefore, the reliability of reception is improved. If the fading is independent across antenna pairs, a maximal diversity gain of n can be achieved: the average error probability can be made to decay as $1/\text{SNR}^{n_R n_T}$ at high SNR, in contrast to the SNR^{-1} for the single-antenna fading channel[4]. More recent work has concentrated on using multiple transmit antennas to achieve diversity[5-9]. In a system with n_T transmit and n_R receive antennas, also assuming independent fading between each antenna pair, the maximum diversity order is $n_R n_T$, corresponding to the number of independently fading paths between the transmitter and receiver.

Consider the use of binary PSK modulation and with a Rayleigh fading channel, we can obtain the relationship of BER, P_b, and diversity order $n_R n_T$ of a MIMO system following[10]:

$$P_b(\gamma_b) = Q(\sqrt{2\gamma_b}) \tag{6.5}$$

where the SNR per bit, γ_b, is given as

$$\gamma_b = \frac{\xi}{N_0}\sum_{j=1}^{n_R}\sum_{i=1}^{n_T} h_{ij}^2 \tag{6.6}$$

where $\frac{\xi}{N_0}\sum_{j=1}^{n_R}\sum_{i=1}^{n_T} h_{ij}^2$ is the instantaneous SNR on the i,jth channel. The probability density function (PDF) $p(\gamma_b)$ is

$$p(\gamma_b) = \frac{1}{(n_R n_T)!\ \bar{\gamma}_c^{-n_R n_T}} \gamma_b^{n_R n_T - 1} e^{-\gamma_b/\bar{\gamma}_c} \tag{6.7}$$

where $\bar{\gamma}_c$ is the average SNR per channel, which is assumed to be identical for all channels. Thus, the integral, P_b, is:

$$P_b = \int_0^{\infty} P_b(\gamma_b) P(\gamma_b) d\gamma_b \tag{6.8}$$

There is a closed-form solution for (6.7), which can be written as[11]

$$P_b = [\frac{1}{2}(1 - \mu)]^{n_R n_T} \sum_{k=0}^{n_R n_T - 1} \binom{n_R n_T - 1 + K}{k} [\frac{1}{2}(1 - \mu)]^k \tag{6.9}$$

where by definition

$$\mu = \sqrt{\frac{\bar{\gamma}_c}{1 + \bar{\gamma}_c}} \tag{6.10}$$

When $\bar{\gamma}_c \gg 1$, $\frac{1}{2}(1 + \mu) \approx 1$ and $\frac{1}{2}(1 - \mu) \approx \frac{1}{4}\bar{\gamma}_c$. Furthermore[10],

$$\sum_{k=1}^{n_R n_T} \binom{n_R n_T - 1 + K}{k} = \binom{2 n_R n_T - 1}{n_R n_T} \tag{6.11}$$

Therefore, when $\bar{\gamma}_c$ is sufficiently large (greater than 10 dB), the probability of error in (6.9) can be approximated as[11]

$$P_b \approx \left(\frac{1}{4\bar{\gamma}_c}\right)^{n_R n_T} \binom{2 n_R n_T - 1}{n_R n_T} \tag{6.12}$$

We observe that the probability of error varies as $1/\bar{\gamma}_c$ raised to the $n_T n_R$th power. Therefore, with diversity, the error rate decreases inversely with the $n_T n_R$th power of the SNR. Therefore, we can obtain the diversity order of a MIMO system by observing the slope of the BER curve in the high-SNR region. Equation (6.9) helps to evaluate the diversity order achieved by MIMO systems.

6.4 Space-time coding

Space-time coding is intended to maximize the diversity gain of a MIMO system and is historically derived from transmit diversity techniques, in which a diversity order up to n_T can be achieved when using n_T transmit antennas and one receive antenna, to mitigate the effects of channel fading through the use of multiple antennas when only one receive antenna is possible.

Two main types of space-time codes exist: space-time trellis codes (STTC)[12,13] and space-time block codes (STBC)[5]. The former are closely related to trellis-coded modulation[6], while the latter is a mapping scheme rather than a true code. We concentrate on describing STBC because they are much simpler to implement, and a simple version of STBC is already incorporated in the UMTS standard[14].

STBC are derived from the mathematical concept of orthogonal designs[6]. In an orthogonal design, symbols are arranged in an array such that all the rows and columns are orthogonal. The block code is constructed in such a way that each column of the array denotes a different transmit antenna, and the rows denote different transmission periods. Complex orthogonal designs are required in wireless systems, that is, the symbols may be complex because they can be taken from the constellation of any suitable modulation scheme.

The simplest form of STBC is the Alamouti scheme[15], the schematic of which is shown in figure 6-3.

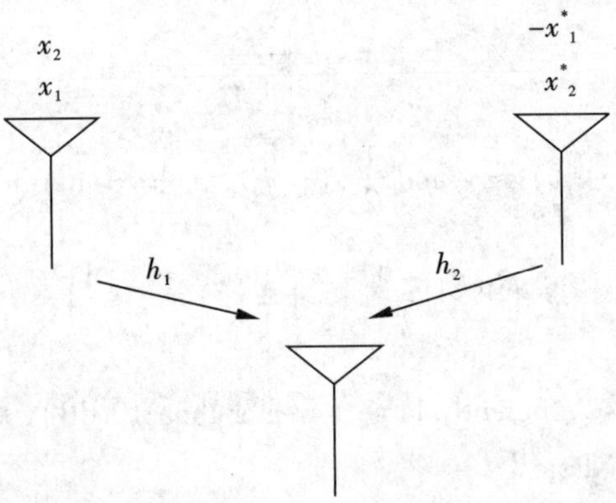

Figure 6-3 Schematic of the Alamouti scheme

The orthogonal design of the code block is:

$$\begin{bmatrix} x_1 & x_2^* \\ x_2 & -x_1^* \end{bmatrix} \tag{6.13}$$

In two successive transmission periods, x_1 and x_2^* are transmitted on one antenna and x_2 and $-x_1^*$ are transmitted on the other.

In this scheme, the decoding technique is quite simple. According to the system in figure 6-3, the received signal for each transmission period is:

$$\begin{aligned} r_1 &= h_1 x_1 + h_2 x_2 + n_1 \\ r_2 &= h_1 x_2^* - h_2 x_1^* + n_2 \end{aligned} \tag{6.14}$$

Based on linear processing at the receiver and following Tarokh's derivation in [5], the decoding algorithm for decoding x_1 and x_2 minimizes the following metrics:

$$|[r_1h_1^*] - (r_2)^*h_2 - x_1|^2 + (-1 + |h_1 + h_2|^2)|x_1|^2 \tag{6.15}$$

$$|[r_1h_1^*] + (r_2)^*h_1 - x_2|^2 + (-1 + |h_1 + h_2|^2)|x_2|^2 \tag{6.16}$$

By assuming that the transmitted symbols have constant power, we can obtain the estimated signals from (6.15) and (6.16) for x_1 and x_2:

$$\hat{x}_1 = r_1h_1^* - (r_2)^*h_2 \tag{6.17}$$

$$\hat{x}_2 = r_1h_2^* - (r_2)^*h_1 \tag{6.18}$$

We can also write these expressions (6.14)-(6.18) in another way:

$$\boldsymbol{r} = \boldsymbol{Cx} + \boldsymbol{n} \tag{6.19}$$

where the new form of the channel matrix $\boldsymbol{C}$ can be written as:

$$\boldsymbol{C} = \begin{bmatrix} h_1 & h_2 \\ -h_2^* & h_1^* \end{bmatrix} \tag{6.20}$$

with $\boldsymbol{x} = \left[x_1, x_2\right]^{\mathrm{T}}$ and $\boldsymbol{n} = \left[n_1, n_2^*\right]^{\mathrm{T}}$. The received signal is now in the form:

$$\boldsymbol{r} = \begin{bmatrix} r_1 \\ r_2^* \end{bmatrix} = \begin{bmatrix} h_1 & h_2 \\ -h_2^* & h_1^* \end{bmatrix} \cdot \begin{bmatrix} x_1 \\ x_2 \end{bmatrix} + \begin{bmatrix} n_1 \\ n_2 \end{bmatrix} \tag{6.21}$$

Therefore, we can obtain

$$r_1 = h_1x_1 + h_2x_2 + n_1 \tag{6.22}$$

$$r_2^* = h_1^*x_2 - h_2^*x_1 + n_2^* \Rightarrow r_2 = h_1x_2^* - h_2x_1^* + n_2 \tag{6.23}$$

which is the same result as in (6.14). The decoding procedure can then be expressed as:

$$\hat{\boldsymbol{x}} = \boldsymbol{C}^{\mathrm{H}}\boldsymbol{r} = \begin{bmatrix} h_1^* & -h_2 \\ h_2^* & h_1 \end{bmatrix} \begin{bmatrix} r_1 \\ r_2^* \end{bmatrix} = \begin{bmatrix} h_1^*r_1 - h_2r_2^* \\ h_2^*r_1 + h_1r_2^* \end{bmatrix} \tag{6.24}$$

which has the same form as that in (6.17) and (6.18). It shows that we can use this simpler form, presented in (6.19)-(6.24), to express the Alamouti STBC.

The simple linear manipulation implies that STBCs are not true codes, and give rise only to a diversity gain, not to a true coding gain. Tarokh *et al.* generalized the Alamouti scheme to an arbitrary number of transmit and receive antennas[5]. The specific case of four transmit antennas for complex symbols, $\mathcal{G}_4$, is used in this chapter, the encoding process of which is:

$$\mathcal{G}_4 = \begin{bmatrix} x_1 & x_2 & x_3 & x_4 \\ -x_2 & x_1 & -x_4 & x_3 \\ -x_3 & x_4 & x_1 & -x_2 \\ -x_4 & -x_3 & x_2 & x_1 \\ x_1^* & x_2^* & x_3^* & x_4^* \\ -x_2^* & x_1^* & -x_4^* & x_3^* \\ -x_3^* & x_4^* & x_1^* & -x_2^* \\ -x_4^* & -x_3^* & x_2^* & x_1^* \end{bmatrix} \tag{6.25}$$

where x_1, x_2, x_3, and x_4 are four complex symbols to be transmitted from 4 transmitted antennas in 8 successive symbol periods. This process achieves full diversity order, but its rate is only 1/2.

Following the derivation in [5] and [16] and generalizing equations (6.19)-(6.24), we can rewrite the received signal for $\mathcal{G}_4$ in the case of multiple receive antennas as:

$$\boldsymbol{r} = \boldsymbol{C}\boldsymbol{x} + \boldsymbol{n} \tag{6.26}$$

where $\boldsymbol{x} = [x_1, x_2, x_3, x_4]^{\mathrm{T}}$ represents the transmitted symbol vector and the channel matrix $\boldsymbol{C}$ is expressed as:

$$\boldsymbol{C} = [\boldsymbol{C}_1, \cdots, \boldsymbol{C}_{n_{\mathrm{R}}}]^{\mathrm{T}} \tag{6.27}$$

where $\boldsymbol{C}_j$ $(j = 1, \cdots, n_{\mathrm{R}})$ is in the form

$$\boldsymbol{C}_j = \begin{bmatrix} \boldsymbol{h}_{12,j} & \boldsymbol{h}_{34,j} \\ \boldsymbol{h}'_{34,j} & -\boldsymbol{h}'_{12,j} \\ \boldsymbol{h}^*_{12,j} & \boldsymbol{h}^*_{34,j} \\ -\boldsymbol{h}'^*_{34,j} & \boldsymbol{h}'^*_{12,j} \end{bmatrix} \tag{6.28}$$

with $\boldsymbol{h}_{mn,j} = \begin{bmatrix} h_{m,j} & h_{n,j} \\ h_{n,j} & -h_{m,j} \end{bmatrix}$, $\boldsymbol{h}'_{mn,j} = \begin{bmatrix} h_{m,j} & -h_{n,j} \\ h_{n,j} & h_{m,j} \end{bmatrix}$.

The received signal and the AWGN vector are written as $\boldsymbol{r} = [\boldsymbol{r}_1, \cdots, \boldsymbol{r}_{n_{\mathrm{R}}}]^{\mathrm{T}}$ with $r_j = [r_{1,j}, \cdots, r_{4,j}, r^*_{5,j}, \cdots, r^*_{8,j}]^{\mathrm{T}}(j = 1, \cdots, n_{\mathrm{R}})$ and $\boldsymbol{n} = [n_1, \cdots, n_{n_{\mathrm{R}}}]^{\mathrm{T}}$ with $\boldsymbol{n}_j = [n_{1,j}, \cdots, n_{4,j}, n^*_{5,j}, \cdots, n^*_{8,j}]^{\mathrm{T}}$.

$h_{i,j}(i = 1,2,3,4; j = 1, \cdots, n_{\mathrm{R}})$ represents the channel between the 4 transmit antennas and n_{R} receive antennas, defined as in equations (6.2)-(6.4). $r_{1,j}, \cdots, r_{8,j}(j = 1, \cdots, n_{\mathrm{R}})$ represent the signals received from n_{R} receive antennas at 8 different symbol periods.

Applying the linear transform to the received signal gives:

$$\hat{\boldsymbol{x}} = \boldsymbol{C}^{\mathrm{H}}\boldsymbol{r} = \boldsymbol{C}^{\mathrm{H}}\boldsymbol{C}\boldsymbol{x} + \boldsymbol{n}' \tag{6.29}$$

By substituting (6.17)-(6.20) into the above equation, we obtain the estimated signal $\hat{x}_1$ of x_1 as:

$$\hat{x}_1 = 2\sum_{j=1}^{n_R}\sum_{i=1}^{n_T}\left|h_{i,j}\right|^2 x_1 + \Xi_1 \tag{6.30}$$

where $\boldsymbol{\Xi} = [\Xi_1, \cdots, \Xi_{n_T}]^T$ and

$$\boldsymbol{\Xi} = \boldsymbol{C}^H \boldsymbol{n} \tag{6.31}$$

Corresponding to $h_{i,j}, i = 1,2,3,4; j = 1,2,\cdots,n_R$;, the random variable Ξ_1 is a zero-mean complex Gaussian random variable with variance $(4/\text{SNR})\sum_{j=1}^{n_R}\sum_{i=1}^{4}\left|h_{i,j}\right|^2$ per real dimension, and the signal power is $4\left[\sum_{j=1}^{n_R}\sum_{i=1}^{4}\left|h_{i,j}\right|^2\right]^2$.

Compare this result with a scenario in which x_1 is transmitted using 1 transmit antenna and $4n_R$ receive antennas. Then, the estimated signal $\hat{x}_1'$ of x_1 is

$$\hat{x}_1' = \sum_{j=1}^{n_R}\sum_{i=1}^{n_T}\left|h_{i,j}\right|^2 x_1 + \Xi_1' \tag{6.32}$$

where Ξ_1' is a zero-mean random complex Gaussian variable with variance $(1/\text{SNR})\sum_{j=1}^{n_R}\sum_{i=1}^{4}\left|h_{i,j}\right|^2$ per real dimension. This result has the same form as that in (6.22), except that the powers of both the signal and noise are reduced by a factor of 1/4.

Therefore, using the encoding form of $\mathcal{G}_4$ provides exactly the same performance as $4n_R$ -level maximum ratio combining.

Figure 6 - 4 illustrates some simulation results of MIMO systems using STBC with quadrature phase shift keying (QPSK) modulation. The simulations are conducted on a frame by frame basis.

The channels are modeled as independent Rayleigh channels, as introduced in equations (6.2) and (6.3), and the noise is modeled as AWGN. The channels are assumed to be quasi-static. We also assume that the channels remain stationary for the duration of several space-time code blocks, which makes it possible for channel estimation to be conducted at the receiver because perfect channel-state information is assumed to be available in the receiver in the decoding algorithm of STBC. The achieved diversity order increases as the number of transmit and receive antennas increases. A MIMO system with 2 transmit and 1 receive antennas (Tx: 2, Rx: 1) using Alamouti's scheme provides a transmission rate of 2(bits/s)/Hz, and for a system with 4 transmit and 1 receive antennas (Tx:4, Rx:1), a transmission rate of 1 (bit/s)/Hz is achieved. A MIMO system with 4 transmit and 2 receive antennas (Tx:4, Rx:2) has the same transmission rate of 1(bit/s)/Hz. The results of the analysis in (6.5)-(6.12) are also presented for comparison. These results give a good

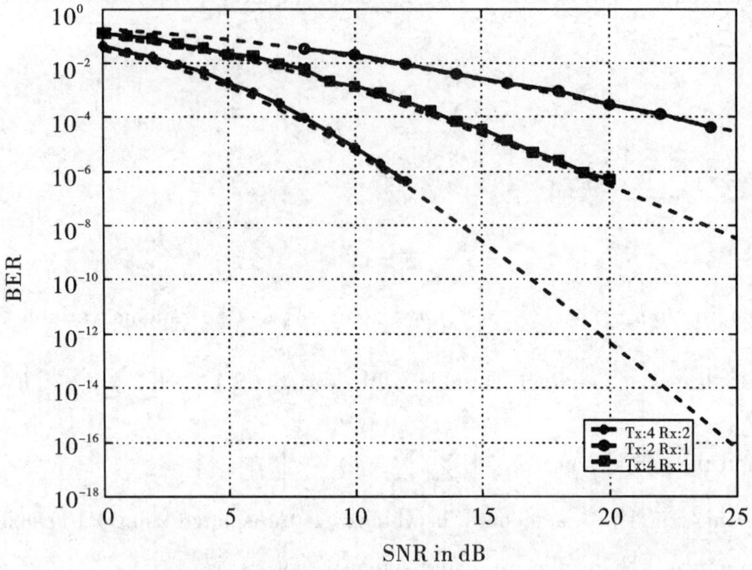

Figure 6 – 4 BER performances of MIMO systems with different numbers of transmit and receive antennas using STBC, solid line-simulation, dashed line-analytical BER of diversity order 2, 4 and 8

match, showing that STBC achieves the full diversity order of $n_R n_T$.

6.5 Spatial multiplexing

In a spatial multiplexing (SM) system, the data stream to be transmitted is first demultiplexed and then sent simultaneously from each transmit antenna. At the receiver, each antenna observes a superposition of the transmitted signals from all the transmit antennas, separates them with appropriate signal processing, and multiplexes them to recover the original data stream. Figure 6-5 illustrates an SM system. The received signals are separated using a multi-antenna detector (MAD), which performs signal processing similar to that used to separate users in a multiuser detector. Forward error correction (FEC) coding is nearly always combined with the SM scheme to enhance performance. The substreams are usually encoded separately, but joint coding may be used.

The most important examples of this SM scheme are the BLAST (Bell labs LAyered Space-Time architecture) techniques, D-BLAST[17] and V-BLAST[18,19] (for diagonal and vertical, respectively). The former uses coding and variable mapping to the antennas, whereas the latter (in principle) does not.

The detection algorithm for SM that yields the optimal performance for minimizing the BER is ML detection. However, the main drawback of this technique is the resulting computational complexity as it has to perform M^{n_T} vector searches, where M is the number of symbols in the

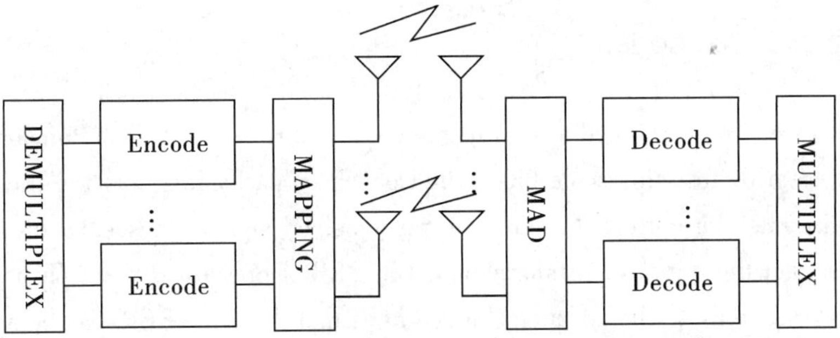

Figure 6-5 Schematic of a spatial multiplexing system

constellation (e. g. , $M = 4$ for QPSK). Therefore, suboptimal techniques, such as the linear processing techniques of zero-forcing (ZF) and MMSE, are considered to reduce the complexity of the detector.

The ZF method provides an estimate of the transmitted symbols as:

$$\hat{\boldsymbol{x}} = \boldsymbol{H}^{\mathrm{H}}(\boldsymbol{H}\boldsymbol{H}^{\mathrm{H}})^{-1}\boldsymbol{r} \tag{6.33}$$

This receiver performs well in the high-SNR regime, but substantial noise enhancement occurs in the low-SNR region.

An alternative linear receiver is the MMSE receiver, which estimates a random variable that minimizes the cost function $E\{|x - \hat{x}|\}$, where $E\{\cdot\}$ stands for the expectation. In this case, an estimate of $\boldsymbol{x}$ is given by:

$$\hat{\boldsymbol{x}} = (\boldsymbol{H}^{\mathrm{H}}\boldsymbol{H} + \sigma^2\boldsymbol{I})^{-1}\boldsymbol{H}^{\mathrm{H}}\boldsymbol{r} \tag{6.34}$$

where $\boldsymbol{I}$ is the identity matrix of size $n_{\mathrm{T}} \times n_{\mathrm{T}}$. In the high-SNR region, the MMSE receiver converges to the ZF receiver. The MMSE receiver is less sensitive to noise at the cost of reduced signal separation quality[2].

Nonlinear techniques also exist, including ordered successive interference cancellation (OSIC), which was the initial decoding algorithm proposed in [18][19]. OSIC detects the received signal from the receive antenna with the highest received SNR, estimates a transmitted symbol, and then cancels this symbol from the received signal to detect the rest of the transmitted symbols. This process is then repeated until all the transmitted symbols have been estimated. This algorithm must be manipulated to output soft symbols[20].

The MIMO system detection for spatial multiplexing is analogous to the multiuser detection as long as we assume that n_{T} is equivalent to the number of users K in the multiuser detection and that n_{R} is equivalent to the spreading factor. The matrix $H^{\mathrm{H}}H$ is considered to be equivalent to the spreading waveform correlation matrix $\boldsymbol{R}$ in multiuser detection.

6.6 Convolutional codes

Forward Error Correction coding is a technique to improve the BER performance of a communication system by inserting some bits, called check or parity bits, into the transmitted data bits. The parity bits can correct the errors in the received signals caused by the channel and help to reconstruct the transmitted information. The BER decreases as the SNR at the input to the demodulator increases. The advantage of coding is that the same BER may be achieved for a lower SNR in a coded system than in a comparable un-coded system[21,22]. FEC techniques are commonly employed together with MIMO-based transmission technologies.

Convolutional coding is a main type of FEC coding, and it is usually described in the form of (l_s, k_s, v_i), where l_s is the length of the current block of coded symbols, k_s is the length of the current block of data symbols, and v_i is the total number of input blocks, called the constraint length. The current block of l_s code symbols depends on the current block of k_s data symbols and on $(v_i - 1)$ previous data blocks. Here, $(v_i - 1)$ is the number of shift registers of the encoder, sometimes called the *memory order* of the code[21]. The code rate is $k_s = l_s$. Note that after all the data symbols are sent through the encoder, $(v_i - 1)$ more zero data bits, called tail bits, are sent to the shift registers of the encoder to ensure that the states of the shift registers are zero after encoding the transmitted data bits.

The encoding operation can be described by the so-called generator polynomials, each representing the generation of an output bit from an input bit[21]. The structure of the (2,1,3) convolutional encoder is depicted in figure 6 – 6, from which we can obtain the generator polynomials for this code as:

$$g^{(1)}(D) = D^2 + D + 1 \tag{6.35}$$

$$g^{(2)}(D) = D^2 + 1 \tag{6.36}$$

where D represents a shift register in the encoder. The highest power of D in the generator polynomials represents the number of shift registers of the encoder, which is 2 in (6.35) and (6.36). The presence of a term in D^i in the polynomials means that there is a connection to a modulo-2 adder for the corresponding code bit from state S_{i-1} of the ith shift register.

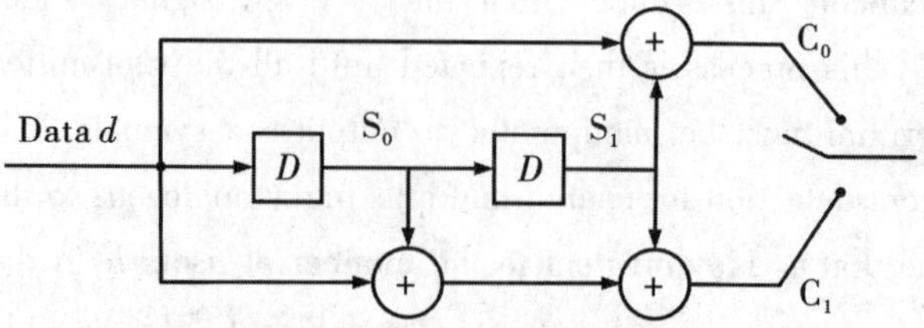

Figure 6–6 Structure of a (2,1,3) convolutional encoder

6.7 Hardware implementation of MIMO and multiuser detectors

Starting from the early work by Foschini, Gans, Teletar, and Paulraj[23,24,11,25], many papers have been published in the area of MIMO-based information theory, algorithms, codes, and so on. Substantial work has focused on the algorithms and protocols that deliver superior BER for a given SNR. Little attention has been given to the real-time implementation of these algorithms. The hardware complexity of the algorithms must be considered so that the MIMO detector can be integrated with the rest of the system. Thus far, MIMO algorithms are generally implemented in DSPs, and it is difficult to achieve high data-rate performance. Field-programmable gate array (FPGA) devices are widely used in signal processing, communications, and network applications because of their re-configurability and support of parallelism.

The FPGA platform has at least three advantages over a DSP processor: the inherent parallelism of the FPGA is equipped for vector processing; reduced instruction overhead; and scalable processing capacity when the FPGA resource is sufficient. The disadvantage is that the development cycle of the FPGA design is usually longer than that of DSP implementation. However, once an efficient architecture is developed and parallel implementation is explored, the FPGA design can greatly improve the processing speed because of its intrinsic density advantage.

In addition, the FPGA platform has several advantages over an ASIC implementation: an FPGA device is reconfigurable to accommodate system configuration changes even during runtime; it has significantly reduced latency compared to that of ASIC; and it is a cost-effective solution. Furthermore, the ASIC implementation is generally applied to a fixed number of antennas and a certain signal constellation. The limitation of an ASIC implementation is the lack of flexibility when the number of antennas or the signal constellation changes.

MIMO detectors are generally implemented on DSPs, such as the BLAST system[26,27]. Because DSPs do not support parallel computation, the implementation speed is often limited, especially when the number of antennas increases. Recently, many ASIC and FPGA implementations of the SD or close-to-ML detectors were reported in [28]-[31]. The VLSI design of depth-first detectors using SD algorithms[32,33] and breadth-first detectors using the M-algorithm[34] have both attracted recent attention[28-30]. For a 4×4 MIMO transmission with 16-QAM, hard-output depth-first detectors[30] achieve much higher throughput than that of their breadth-first counterparts[28,29]. In general, the average computational complexity of depth-first hard-output detection is lower than that of its breadth-first counterpart due to its ability to adaptively tighten the search radius constraint. For soft-output detection, it is not trivial to adaptively change the search radius for both depth-first search and breadth-first search, and breadth-first search has the advantage that it can naturally generate an ordered candidates list for *a posteriori probability* (APP) calculation[3].

Table 6-1 Comparison of the implementations of 4×4 16 QAM MIMO detection

Techniques	BER	Hardware platform	Clock frequency	Throughput at SNR=20 dB(Mbps)	FPGA no. slices or gate ASIC
K-best 1[35,28]	close ML	ASIC	100 MHz	10	52000
K-best 2[35,29]	close ML	ASIC	100 MHz	52	91000
Depth-first SD1 (ℓ^2-norm)[31,30]	ML	ASIC	51 MHz	73	117000
Depth-first SD2 (ℓ^1-norm)[31,30]	close ML	ASIC	71 MHz	169	50000
Depth-first SD3[31]	ML	FPGA	50 MHz	114.5	21467

Table 6-1 gives an overview and comparison of the relevant hardware implementations of 4×4 16-QAM MIMO detection algorithms[30]. Currently, most hardware implementations of MIMO communications apply to small systems. The depth-first SD1 and SD2 are based on the ASIC implementation. The SD2 implementation has twice the throughput as that of the SD1 in the half-chip area. Depth-first tree transversal is implemented in a sequential and non-pipelined manner, whereas the *K*-best algorithm is based on a parallel and pipelined hardware structure with reduced chip area. In addition, the *K*-best approach guarantees constant throughput at the expense of a performance loss. The average throughput of the depth-first approach can match that of the *K*-best approach, but the throughput of the worst-case scenario may decrease substantially. The FPGA SD3[30] has a similar performance to that of the SD1; however, its complexity is significantly reduced. Moreover, these real-time implementations of SD in table 6-1 do not include channel matrix preprocessing, such as QR decomposition and Cholesky factorization, which consume considerable hardware resources.

Multiuser detector implementation is more difficult because it is mainly applied to large systems. Some related references are available, e. g. , the FPGA implementation of an adaptive MMSE algorithm presented in [35] and the FPGA multiuser detector based on a cascade of adaptive filters for asynchronous WCDMA systems presented in [36].

Exercises

1 Explanation of glossary

1-1 MIMO

1-2 FEC

1-3 Diversity

2 Choice question

2-1 Space-Time codes include.

A. Alamouti Code B. MMSE

C. Convolutional code D. ASIC

2-2 There are nonlinear techniques such as ( ).

A. DCD B. OSIC

C. MMSE D. ZF

3 Short answer question

3-1 Analyze the principle of spatial multiplexing system.

3-2 Explain that how forward error correction serves for the communication.

3-3 How to choose the proper hardware platform (DSP, FPGA, ASIC) to implement algorithm?

Reference

[1] YU D. Multiuser Detection in Multiple Input Multiple Output Orthogonal Frequency Division Multiplexing Systems by Blind Signal Separation Techniques. Florida International University, 2012[2018-08-25]. http://digitalcommons. fiu. edu/cgi/viewcontent. cgi? article=1737&context=e.

[2] VCELAK J, JAVORNIK T, SYKORAJ, et al. Multiple-Input Multiple-Output Wireless Systems. Electrontechnical Review, 2003, 70(4): 234-239.

[3] GESBERT D, SHAFI M, SHIUD, et al. From theory to practice: An overview of MIMO space-time coded wireless systems. In IEEE Journal on Selected Areas in Communications, April 2003, 21(3): 281-302.

[4] ZHENG L Z, TSE D N C. Diversity and multiplexing: a fundamental tradeoff in multiple-antenna channels. IEEE transactions on Information Theory, May 2003, 49(5): 1073 - 1096.

[5] TAROKH V, JAFARKHANI H, CALDERBANK A R. Space-time block coding for

wireless communications: performance results. IEEE Journal on Selected Areas in Communications, Mar. 1999, 17(3): 451-460.

[6] BIGLIERI E, DIVSALAR D, MCLANE PJ, et al. Introduction to trellis coded modulation with applications. New York: Macmillan, 1991.

[7] GUEY J, FITZ M, BELLM, et al. Signal design for transmitter diversity wireless communication systems over Rayleigh fading channels. In Pro. IEEE Vehicular Technology Conference, 1996: 136-140.

[8] TAROKH V, JAFARKHANI H, CALDERBANK A R. Space-time block codes from orthogonal designs. IEEE transactions on Information Theory, July 1999, 45: 1456-1467.

[9] MARZETTA T L, HOCHWALD B M. Capacity of a mobile multiple-antenna communication link in Rayleigh flat fading. IEEE transactions on Information Theory, Jan 1999, 49: 139-157.

[10] PROAKIS J G. Digital Communications (3rd edition). New York: McGraw-Hill International Editions, 1995.

[11] PAULRAJ A J, PAPADIAS C B. Space-time processing for wireless communications. IEEE Signal Processing Magazine, 1997, 14(6): 49-83.

[12] TAROKH V, SESHADRI N, CALDERBANK A R. Space-time codes for high data rate wireless communication: performance criterion and code construction. IEEE Journal on Selected Areas in Communications, Mar. 1998, 44(2): 744-765.

[13] NAGUIB A F, TAROKH V, SESHADRIN, et al. A space-time coding modem for high-data-rate wireless communications. IEEE Journal on Selected Areas in Communications, Oct. 1998, 16(8): 1459-1478.

[14] SANDHU S, PAULRAJ A. Space-time block codes: A capacity perspective. IEEE Communications Letters, Dec. 2000, 4(12): 384-6.

[15] ALAMOUTI S M. A simple transmit diversity technique for wireless communications. IEEE Journal on Selected Areas in Communications, Oct 1998, 16(8): 1451-1458,.

[16] ZHENG F C, BURR A G. Signal detection for orthogonal space-time block coding over time-selective fading channels: a PIC approach for the G_i systems. IEEE transactions on Communications, June 2005, 53(6): 969-972.

[17] FOSCHINI G J. Layered space-time architecture wireless communication in a fading environment when using multi-element antennas. Bell labs Tech Journal, autumn 1996, 1(2): 41-59.

[18] GOLDEN G D, FOSCHINI G J, VALENZUELA RA, et al. Detection algorithm and initial laboratory results using V-BLAST space-time communication structure. Electronic Letter, Jan 1999, 35(1): 14-16.

[19] GOLDEN G D, FOSCHINI G J, VALENZUELA RA, et al. Simplified processing for high spectral efficiency wireless communication employing multi-element arrays. IEEE Journal on Selected Areas in Communications, Nov 1999, 17(11): 1841-1852.

[20] LARSSON E G, EDFORS O, TUFVESSON F, et al. Massive MIMO for next generation wireless systems. IEEE communications magazine, Feb. 201452(2):186-95.

[21] BURR A G. Modulation and Coding for Wireless Communications, Prentice Hall, 2001.

[22] SWEENEYP. Error Control Coding: An Introduction, Prentice Hall, 2001.

[23] FOSCHINI G, GANS M. On limits of wireless communications in a fading environment when using multipleantennas. Wireless Personal Communications, Mar. 1998, 6 (3):311-335.

[24] TELATAR I. Capacity of multi-antenna Gaussian channels. European Transactions onTelecommunications, 1999, 10(6):585-595.

[25] PAULRAJ A J, GORE D A. An overview of MIMO communications-a key to gigabit wireless. Proceedings of the IEEE, 2004, 92(2):198-217.

[26] ADJOUDANI A, BECK E C, BURG AP, et al. Prototype experience for MIMO Blast over third-generation wireless system. IEEE J. Sel. Areas Commun., Mar. 2003, 21(3): 440-451.

[27] GUILLAUD M, BURG A, RUPPM, et al. Rapid prototyping design of a 4x4 Blast-over-UMTS system. In Proc. 35th Asilomar Conf. Pacific Grove, CA., Nov. 2001, 2: 1256-1260.

[28] WONG K, TSIU C, CHENG R K, et al. A VLSI architecture of a K-best lattice decoding algorithm for MIMO systems. In Proc. IEEE International Symposium on Circuits and Systems (ISCAS'02), Scottsdale, AZ, May 2002, 3:273-276.

[29] GUO Z, NILSSON P. A VLSI architecture of theSchnorr-Eurchner decoder for MIMO systems. In Proc. IEEE 6th Circuit and Systems Symposium on Emerging Technologies: Frontiers of Mobile and Wireless Communication, Shanghai, China, June 2004, 1:65-68.

[30] BURG A, BORGMANN M, WENKM, et al. VLSI Implementation of MIMO detection using the sphere decoding algorithm. IEEE J. Solid-State Circuits, Jul. 2005, 40(7): 1566-1577.

[31] BARBERO L G, THOMPSON J S. Rapid prototyping of a fixed-throughput sphere decoder for MIMO systems. In Proc. IEEE ICC'06, Istanbul, Turkey, June 2006, 7: 3082-3087.

[32] FINCKE U, POHST M. Improved methods for calculating vectors of short length in a lattice, including a complexity analysis. Math. Comput., Apr. 1985, 44(170):463-471.

[33] HOCHWALD B M, TEN BRINK S. Achieving near-capacity on a multiple-antenna channel. IEEE Trans. Commun., Mar. 2003, 51(3):389-399.

[34] ANDERSON J B, MOHAN S. Sequential coding algorithms: A survey and cost analysis. IEEE Trans. Commun., Feb. 1984, 32(2):169-176.

[35] HO Q T, MASSICOTTE D. FPGA implementation of adaptive multiuser detector for DS-CDMA systems. In Proceedings of 14th International Conference on Field Programmable

Logic and Application (FPL'04), Aug. -Sep. 2004, 1:959-964.

[36] HO Q T, MASSICOTTE D. FPGA implementation of an MUD based on cascade filters for a WCDMA system. EURASIP Journal on Applied Signal Processing, 2006, 1:1-12.

Chapter 7 Box-constrained DCD-based multiuser detector

7.1 Introduction

In CDMA systems, multiuser detection is capable of providing a high detection performance[1]. The use of multiuser detection significantly increases the spectral efficiency of wireless communication systems. This technology has developed into an important field of multi-access communications. The conventional detector (matched filter) performs poorly in situations where the energies of the multiuser signals are different or when there are many users. The exponential computational complexity of the optimal ML detector makes it infeasible in real-time operations. The SD algorithm has been proven to simplify ML multiuser detection[2]. However, the hardware implementation of matrix factorization, such as Cholesky decomposition or QR decomposition, is challenging[3,4]. Therefore, the SD algorithm is suitable only for small systems.

The box-constrained dichotomous coordinate descent (DCD) algorithm provides a multiplication and division-free solution to the normal equations, which makes it more suitable for real-time implementation. The box-constrained DCD algorithm[5] is studied in this chapter. The box-constrained DCD-based multiuser detector is proposed to solve large systems. The performance of the box-constrained DCD-based multiuser detector applied in systems with a large number of users is presented.

The box-constrained DCD algorithm is implemented on an FPGA board. Two FPGA architecture designs are described in this chapter. The serial implementation of the box-constrained DCD algorithm has much less complexity than that of the FPGA implementation of the sphere decoders because the algorithm is free of explicit multiplications and divisions and requires only addition and bit-shift operations. The parallel architecture design can improve the data throughput relative to that of the serial-based design. Therefore, a parallel FPGA design of the box-constrained DCD detector is also presented.

7.2 Formulation of the multiuser detection problem

We consider a K-user synchronous CDMA system using BPSK modulation in an AWGN channel. The matched filter output in the receiver is given by:

$$\boldsymbol{\theta} = \boldsymbol{R}\boldsymbol{x} + \boldsymbol{n} \tag{7.1}$$

where the vector $\boldsymbol{x} \in \{-1, +1\}^K$ contains the bits transmitted by K users, $\boldsymbol{R}$ is a real-valued $K \times K$ matrix, $\boldsymbol{x}$ and $\boldsymbol{\theta}$ are real-valued $K \times 1$ vectors and $\boldsymbol{n}$ is a real-valued zero-mean Gaussian random vector with covariance matrix $\sigma^2\boldsymbol{R}$. The optimal ML multiuser detector estimates the vector $\boldsymbol{x}$ by minimizing the following quadratic function with an integer constraint:

$$\hat{\boldsymbol{x}} = \underset{\boldsymbol{x} \in \{-1,+1\}^K}{\operatorname{argmin}} \left\{ \frac{1}{2}\boldsymbol{x}^{\mathrm{T}}\boldsymbol{R}\boldsymbol{x} - \boldsymbol{\theta}^{\mathrm{T}}\boldsymbol{x} \right\}. \tag{7.2}$$

Although the ML detector provides the best detection performance, it is not practical due to its high complexity. The sphere decoder can provide performance identical to that of the ML detector with significantly reduced average complexity. However, at low SNRs, the worst-case complexity of the sphere decoder is exponentially proportional to the number of users, which prevents the use of the sphere decoder in systems with a large number of users[6].

7.3 Box-constrained DCD algorithm

The box-constrained DCD-based multiuser detector[5] uses the box constraint $\boldsymbol{x} \in [-1, +1]^K$ in the quadratic minimization (7.2). Table 7-1 presents the box-constrained DCD algorithm. The box-constrained DCD algorithm is designed to offer a simple solution for the vector $\boldsymbol{x}$ without explicit multiplication and division. The final accuracy of the solution vector $\boldsymbol{x}$ depends on the number of bits (M_b), the number of iterations, and the conditional number of the system matrix. The first set of iterations in the algorithm determines the most significant bit ($m = 1$) for all elements of $\boldsymbol{x}$ using a step size parameter d. The subsequent sets of iterations determine the less significant bits up to a suitable number of bits (a maximum of M_b). The residual vector is given by $\boldsymbol{r} = \boldsymbol{\theta} - \boldsymbol{R}\bar{\boldsymbol{x}}$, where $\bar{\boldsymbol{x}}$ is the initialization of $\boldsymbol{x}$. In this chapter, $\bar{\boldsymbol{x}}$ is set to zero, and $\boldsymbol{r}$ is set equal to $\boldsymbol{\theta}$. The step size d is reduced by a power of two at step 1, so no explicit multiplication or division is conducted because all the multiplication and division operations can be replaced by simple bit shifts. If an element of the solution vector is updated in an iteration, the iteration is labeled "successful". For every step size update, the algorithm repeats the successful iterations until all the elements of residual vector $\boldsymbol{r}$ become so small that the condition at step 4 is not met for all j or until $\boldsymbol{x}$ overflows the range $[-H, +H]$ in step 6, where $H = 1$ for BPSK modulation. The computational load of the algorithm depends mainly on the number of successful iterations N_u and the number of bits M_b. A limit for the number of successful iterations N_u can be predefined and used as a stopping condition (at step 10). If no such limit is defined or if the limit is sufficiently high, the accuracy of the solution is 2^{-M_b+1}.

Table 7-1 Box-constrained DCD algorithm

Initialization: $\boldsymbol{x}=\bar{\boldsymbol{x}}, \boldsymbol{r}=\boldsymbol{\theta}-\boldsymbol{R}\bar{\boldsymbol{x}}, H=1, p=0.$

for $m=1:M_b$

1. $d=2^{-m+1}$
2. **Flag** $=0$ …pass
3. **for** $j=1:K$ …iteration
4. **if** $|r(j)|>(d/2)R(j,j)$ **then**
5. $x=x(j)+\text{sign}(r(j))\cdot d$
6. **if** $|x|<H$ then
7. $x(j)=x$
8. $r=r-\text{sign}(r(i))\cdot d\cdot \boldsymbol{R}(:,j)$
9. **Flag** $=1, p=p+1$
10. **if** $p>N_u$ **end algorithm**
11. **end** j**-loop**
12. **if Flag** $=1$, go to 2

end m-loop

A successful iteration requires one addition for comparison (at step 4) and $(K+1)$ additions for updating the residual vector $\boldsymbol{r}$ and element $x(j)$. For an unsuccessful iteration, only one addition is used for the comparison. The worst-case complexity corresponds to the unlikely situation where only the last mth bit has N_u successful iterations. This means that the calculations of the first (M_b-1) bits do not contain any successful iterations and therefore require $(M_b-1)K$ additions. The worst-case complexity for calculating the last bit (corresponding to $m=M_b$) occurs when only one successful iteration occurs among the K iterations $(j=1,\cdots,K)$. This situation requires K additions for the comparison and $(K+1)$ additions to update the residual vector $\boldsymbol{r}$ (at step 8) and element $x(j)$ (at step 5). In total, N_u successful iterations require $N_u(2K+1)$ additions.

Therefore, the complexity of the box-constrained DCD algorithm isupper bounded by $K(2N_u+M_b-1)+N_u$ additions. However, in a typical situation, there should be several successful iterations in each pass, and the average complexity will be close to $2KN_u$.

Figure 7-1 shows a block diagram of the DCD processor.

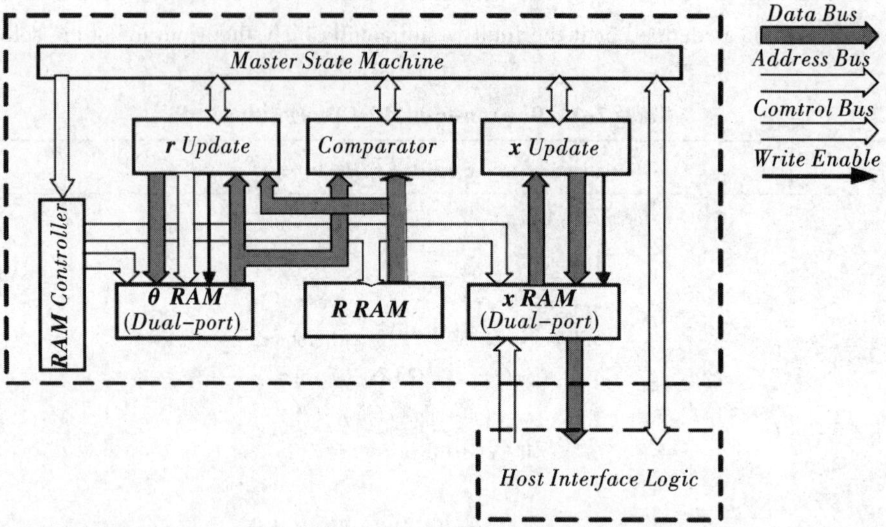

Figure 7-1 DCD processor block diagram

7.4 Fixed-point serial implementation of the box-constrained DCD algorithm

The fixed-point box-constrained DCD algorithm is implemented directly in VHDL on an FPGA platform. The development board is a Xilinx Virtex-II Pro Development System[7] with an XC2VP30 FPGA chip (FFT896 package, speed grade 7). The fixed-point DCD algorithm is synthesized and downloaded to this FPGA chip through Xilinx ISE 8.1i running at a clock frequency of 100 MHz. The design uses 16-bit Q15 number format to represent elements of the matrix $\boldsymbol{R}$. To avoid overflow errors, 32-bit fixed-point integers are used to represent the elements in vectors $\boldsymbol{\theta}$ and $\boldsymbol{x}$. These elements are a block diagram of the DCD processor block diagram of the DCD processor limited to the range $[-2^{16}, 2^{16})$. We treat the vector stored in θ RAM as the residual vector $\boldsymbol{r}$.

Table 7-2 describes the real-valued serial implementation of the fixed-point box-constrained DCD algorithm. There are six states of the fixed-point box-constrained DCD algorithm, as shown in figure 7-2. Figure 7-1 shows a block diagram of the DCD processor. The transitions among these states in the fixed-point box-constrained DCD algorithm are presented in figure 7-2. The vectors $\boldsymbol{x}$ and $\boldsymbol{r}$ are stored in $\boldsymbol{x}$ RAM and $\boldsymbol{\theta}$ RAM, respectively. Vectors $\boldsymbol{x}$ and $\boldsymbol{r}$, bit counter m, successful iteration counter p, pre-scaling counter Δm, element index j and Flag are initialized to state 0. In state 1, if $m \neq 0$, the algorithm chooses step size $d = 2^m$, decrements the bit counter m by one and increments the pre-scaling factor Δm by one. If the least significant bit of the solution is achieved ($m = 0$), the algorithm stops.

Table 7-2 Fixed-point real-valued box-constrained DCD algorithm

state	Operation	Cycles
0	Initialization: $\boldsymbol{x} = 0, \boldsymbol{r} = \boldsymbol{0}$, $m = M_b$, $p=0$ $\Delta m=0, j=1$, Flag=0	
1	if $m=0$, algorithm stops else, $m=m-1, d=2^m$, $\Delta m=\Delta m+1$	1
2	$c = R(j,j)/2 - \vert r(j) \vert \times 2^{\Delta m}$ $x_t = x(j) + \text{sign}(r(j)) \cdot d$ $\vert x_t \vert \leqslant H$, then $\alpha=0$; else $\alpha=1$	1
3	If $c<0$ and $\alpha=0$, then goto state 4 else, goto state 5	1
4	$x(j) = x_t$ $\boldsymbol{r} = \boldsymbol{r} \times 2^{\Delta m} - \text{sign}(r(j)) \cdot \boldsymbol{R}(:,j)$ $\Delta m=0, p=p+1$, Flag=1 If $p = N_u$, algorithm stops	K
5	$j = (j)\text{mod}(K) + 1$ if $j=1$, and Flag=1 then Flag=0 goto state 2 else if $j=1$ and Flag=0, then goto state 1 else, goto state 2	1
Total	$\leqslant 4KN_u + 3K(M_b - 1) + M_b$	

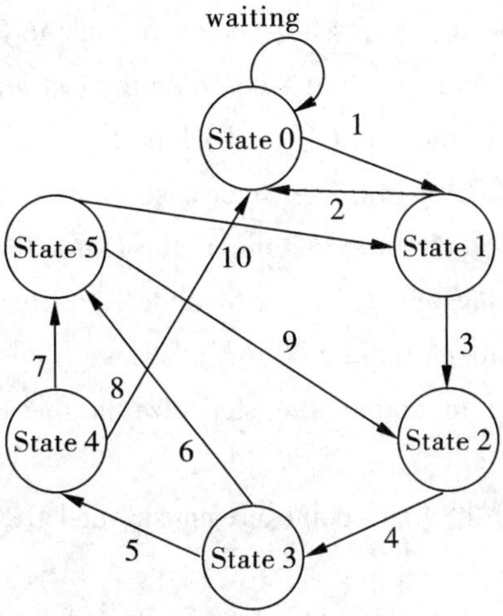

Line 1:Start computing
Line 2:m loop done
Line 3:do comparison
Line 4:go to check the comparison result
Line 5:Successful iteration,goto do update.
Line 6:Unsuccessful iteration,goto do next element comparison.
Line 7:to update counter j.
Line 8:Maximum iteration meets,exit the computing
Line 9:Flag=0,go back to next comparison
Line 10:First element is unsuccessful iteration, update step size and m.

State 0:Initialization state.
State 1:m-loop control & step size updating.
State 2:Comparison.
State 3:Judge "successful" or "unsucessful".
State 4:$x(j)$ update and vector $\boldsymbol{r}$ update
State 5:j loop control.

Figure 7-2 DCD Master State Machine

In state 2, two cycles are needed before each comparison because of a latency delay when reading from the RAM. The RAM Controller asserts the addressesof $r(j)$, $R(j,j)$ and $x(j)$, which are stored in $\boldsymbol{\theta}$ RAM, $\boldsymbol{R}$ RAM and $\boldsymbol{x}$ RAM, respectively. Simultaneously, $r(j)$ is scaled by $2^{\Delta m}$ by bit shifting. In addition, the $\boldsymbol{x}$ Update Logic reads $x(j)$ from $\boldsymbol{x}$ RAM, pre-updates $x(j)$ and stores the updated element in register x_t. Moreover, it checks whether x_t is in the range $[-H, H]$ ($\alpha = 0$) or not ($\alpha = 1$) before proceeding to state 3.

In state 3, the Master State Machine checks the result of the comparison from state 2 to determine whether the iteration is "successful". If the iteration is "successful", the algorithm goes to state 4 to update $\boldsymbol{x}$ and $\boldsymbol{r}$; otherwise, it proceeds to state 5.

In state 4, the $\boldsymbol{r}$ Update Logic updates all the elements of vector $\boldsymbol{r}$. The RAM Controller indicates the addresses of the elements of column $\boldsymbol{R}(:,j)$ and vector $\boldsymbol{r}$. The corresponding elements of $\boldsymbol{r}$ and $\boldsymbol{R}(:,j)$ are loaded into the $\boldsymbol{r}$ Update Logic. $x(j)$ is updated by directly copying the value from register x_t. Meanwhile, Δm is cleared to 0, the iteration counter p is incremented by one, and the binary variable Flag is set to 1, which indicates that the current iteration is "successful". Then, it is checked whether "successful" iteration counter p reaches the maximum number N_u. If p is less than N_u, the Master State Machine proceeds to state 5; otherwise, it stops.

State 5 first updates index j. Then, it decides whether to update the Flag and which state to proceed to depending on j and Flag.

The number of clock cycles required for each state is shown in table 7-2. When using pipeline technology, state 4 requires K cycles to update all the elements in vector $\boldsymbol{r}$ and the element in $\boldsymbol{x}$. Other states require only one cycle to execute. State 2 also requires 2 additional clock cycles because there is a latency when reading data from RAM. The total cycles required mainly depends on the number of successful update iterations and the number of bits. The upper number of cycles in table 7-2 can be considered to be the worst-case complexity of the fixed-point box-constrained DCD algorithm. The worst-case situation occurs when only the last mth bit has N_u successful iterations. Therefore, the calculations of the first $(M_b - 1)$ bits do not contain any successful iterations and hence require $(M_b - 1)3K$ cycles. The worst case for calculating the last bit ($m = 1$) occurs when only one successful iteration exists among the K iterations ($j = 1, \cdots, K$). This requires $3K$ cycles for the comparison and K cycles to update the residual vector $\boldsymbol{r}$ and element $x(j)$. In total, N_u successful iterations require $N_u(4K)$ cycles. In addition to the above required cycles, M_b cycles are required to update the step size in the whole process.

Therefore, the maximum number of clock cycles of the fixed-point box-constrained DCD algorithm is $4KN_u + 3K(M_b - 1) + M_b$.

Figure 7-1 presents the FPGA architecture of the box-constrained DCD algorithm, which includes sub-modules: Master State Machine, RAM Controller, Comparator, $\boldsymbol{r}$ Update Logic, and $\boldsymbol{x}$ Update Logic[8]. The matrix $\boldsymbol{R}$, vector $\boldsymbol{r}$, and vector $\boldsymbol{x}$ are saved in $\boldsymbol{R}$ RAM, $\boldsymbol{\theta}$ RAM and $\boldsymbol{x}$

RAM, respectively.

Master State Machine: The Master State Machine is required to track the current iteration number, select and update the step size and decide which state to execute.

RAM Controller: The RAM Controller drives one of the address ports of $\boldsymbol{\theta}$ RAM and $\boldsymbol{R}$ RAM and both address ports of $\boldsymbol{x}$ RAM. During the "Comparison" step in an iteration, the RAM Controller provides the addresses of $R(j,j)$, $r(j)$, and $x(j)$. The update operation is conditional; if the iteration is "successful", then the RAM Controller sequentially increments the addresses in vector $\boldsymbol{r}$ and matrix $\boldsymbol{R}$ so that all K elements in the vector are updated. Meanwhile, the RAM Controller indicates the address of $x(j)$ in $\boldsymbol{x}$ RAM to update element $x(j)$. The RAM Controller then indicates the addresses of $r(j+1)$ and $R(j+1,j+1)$ for the next comparison. However, if the iteration is "unsuccessful", the RAM Controller immediately indicates the addresses of $r(j+1)$ and $R(j+1,j+1)$ (without update) for the next comparison.

Comparator: Figure 7-3 shows the Comparator architecture used for comparing $r(j)$ and $(d/2)R(j,j)$ in step 4 of table 7-1. The Comparator also scales $r(j)$ to compensate for d in state 2 of table 7-2 and passes the sign bit of the result c to the Master State Machine.

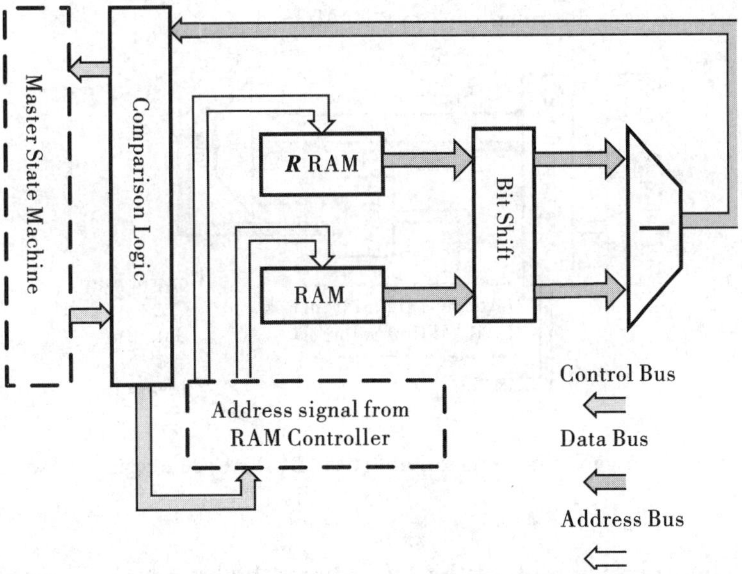

Figure 7-3 Comparator architecture

***r* Update Logic**: Figure 7-4 presents the architecture of the $\boldsymbol{r}$ Update Logic. The initialization vectors $\boldsymbol{r}$ and $\boldsymbol{x}$ are stored concurrently in $\boldsymbol{\theta}$ RAM and $\boldsymbol{x}$ RAM, respectively. The $\boldsymbol{\theta}$ RAM is filled with scaled data, while the $\boldsymbol{x}$ RAM is cleared to zeros. Each element is read from $\boldsymbol{\theta}$ RAM, updated as $\boldsymbol{r} = \boldsymbol{r} \times 2^{\Delta m} - \text{sign}(r(j)) \cdot \boldsymbol{R}(:,j)$, and written back to $\boldsymbol{\theta}$ RAM.

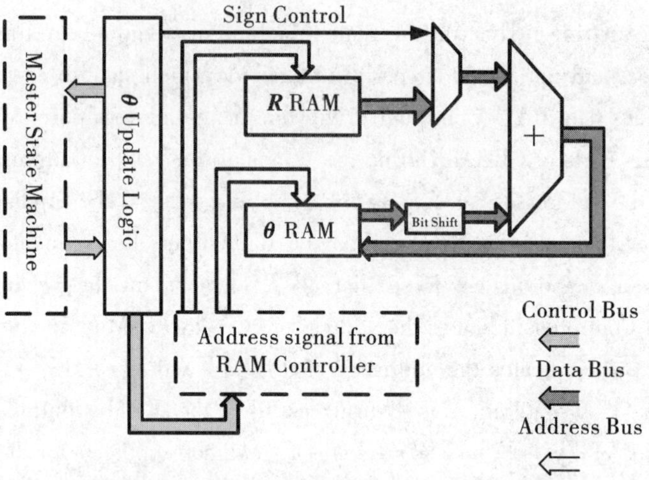

Figure 7–4 Architecture of DCD *r* Update logic

x* Update Logic:** Figure 7–5 shows the ***x Update Logic. The ***x*** Update Logic reads element $x(j)$ from ***x*** RAM. $x(j)$ is updated by adding or subtracting the step size d according to the sign of $r(j)$. The updated $x(j)$ is written back to ***x*** RAM.

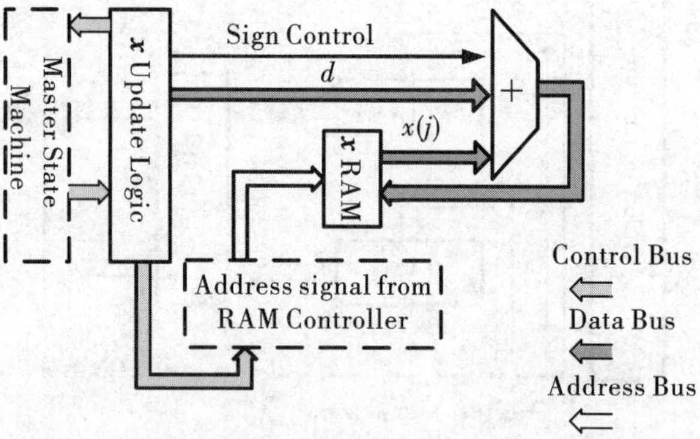

Figure 7–5 Architecture of the DCD *x* Update logic

7.4.1 Detection performance of the DCD-based box-constrained detector

The BER performance of the box-constrained DCD detector is evaluated using 10^5 simulation trials. We assume the simulation scenario is an AWGN channel with perfect power control, where the users employ randomly generated spreading sequences. We consider a high loaded scenario, for which the number of users $K = 50$ and the spreading factor $SF = 53$, and a less loaded scenario, for which $K = 20$ and $SF = 31$. We also investigate the detector using both fixed-point and floating-point implementations for these scenarios. The floating-point implementation results are obtained from MATLAB, and the fixed-point implementation results are obtained from the FPGA board.

Figure 7-6 shows the BER performance as a function of SNR for a highly loaded scenario and different M_b. There is no distinguishable difference between the fixed-point FPGA results and the floating-point results. Even the box-constrained DCD detector with two-bit representation ($M_b = 2$) outperforms the MMSE detector over the SNR range from 0 dB to 17.5 dB. Furthermore, when BER $= 10^{-3}$, the box-constrained DCD detector with $M_b = 4$ achieves a 7.5 dB improvement in the BER performance compared with that of the MMSE detector. When $M_b = 14$, the BER performance is not greatly improved in comparison with that when $M_b = 4$, indicating that a further increase in the number of bits M_b will not result in substantial improvement in the BER performance.

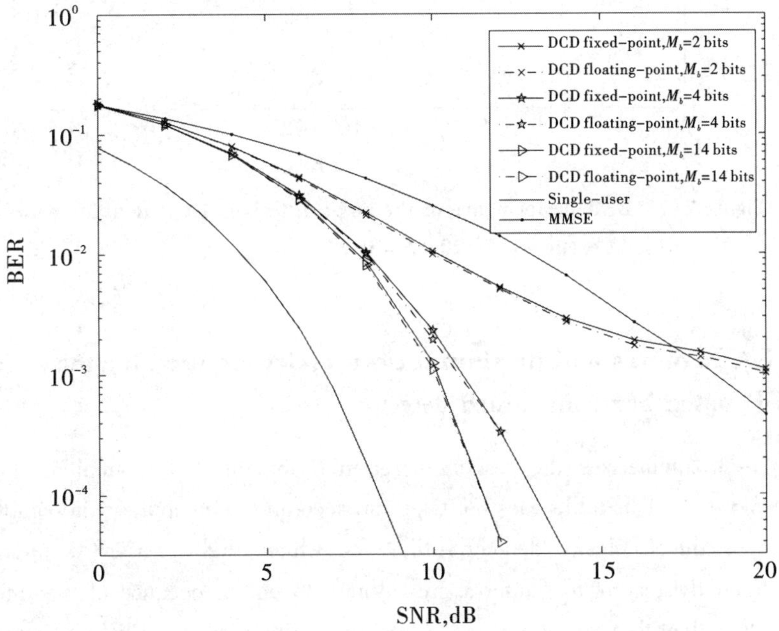

Figure 7 - 6 BER performance of the DCD-based box-constrained multiuser detector for different numbers of bits M_b in a highly loaded multiuser scenario; $K=50$, $SF=53$

Figure 7-7 shows the BER performance for a less loaded scenario. The parameter M_b in the box-constrained DCD algorithm can be chosen as any integer to represent the estimate data. In this simulation, we use M_b values of 2, 4, 14 to assess the detection performance. The BER performance of the FPGA fixed-point implementation of the box-constrained DCD detector is close to that of the floating-point solution. Even the box-constrained DCD detector with two-bit representation ($M_b = 2$) outperforms the MMSE detector over the SNR range from 0 dB to 20 dB. Moreover, figure 7-7 also shows that the detection performance of the box-constrained DCD detector with $M_b = 4$ is similar to that with $M_b = 14$. Therefore, $M_b = 4$ is sufficient to implement the box-constrained DCD detector.

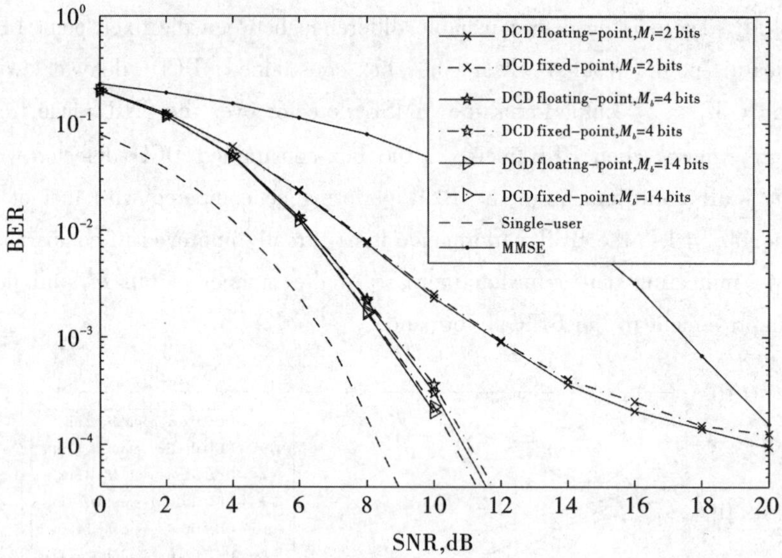

Figure 7-7 BER performance of the box-constrained DCD detector with different M_b; $K=20$, $SF=31$.

7.4.2 FPGA resources and maximum clock cycles required for the serial DCD-based box-constrained detector

Table 7 - 3 summarizes the resources required for the FPGA implementation of SD algorithm for $K=4$ [2]. This table does not take into account the computational complexity of the pseudoinverse or the Cholesky decomposition, for which the hardware implementation is complex. Additionally, results for larger K are difficult to obtain because of the infeasibility of implementing this algorithm for more users. Table 7-4 summarizes the FPGA resources required for the DCD box-constrained algorithm for $K = 50$ and $M_b = 4$. This algorithm uses a much larger number of users than that used for the SD algorithm and is therefore not an equivalent comparison; however, it is still a worthwhile comparison since the sphere detector can provide optimal detection performance. The number of multipliers used in the box-constrained DCD algorithm implementation is zero because multiplication operations are achieved by simple bit-shift operations. The number of slices used in the SD algorithm implementation is approximately 33 times greater than that used in the box-constrained DCD algorithm. Furthermore, the number of RAMs used in the box-constrained DCD detector implementation is approximately 20 times less than that used in the sphere decoder. We have also implemented the box-constrained DCD detector when $K = 4$ (not presented here) and have found that the number of slices used in the box-constrained DCD algorithm does not vary significantly with variation in the system size, i. e., K from 4 to 50. However, the RAM storage for the equation operands increases as the system size increases.

Table 7-3 FPGA resources needed for the SD algorithm[2] for $K = 4$

FPGA Resources	Total available	Usage
Slices	33088	12721
Multipliers	328	160
Block RAMs	328	82

Table 7-4 FPGA resources needed for the DCD box-constrained algorithm for $K = 50, M_b = 4$

FPGA Resources	Total available	Usage
Slices	13696	387
Multipliers	136	0
Block RAMs	136	4

Table 7-5 shows the maximum number of clock cycles of the box-constrained DCD algorithm for different K and M_b. For the same K, the maximum number of clock cycles required increases as M_b increases. For the same M_b, the maximum number of clock cycles needed increases as K increases. In conclusion, the update time increases as either M_b or K increases.

Table 7-5 Worst-case number of clock cycles required for the box-constrained DCD algorithm for different M_b

	$K=4$	$K=20$	$K=50$	$K=110$
$M_b=2$	$16N_u + 14$	$80N_u + 62$	$200N_u + 152$	$440N_u + 332$
$M_b=4$	$16N_u + 40$	$200N_u + 454$	$200N_u + 454$	$440N_u + 994$
$M_b=14$	$16N_u + 170$	$200N_u + 1964$	$200N_u + 1964$	$440N_u + 4304$

7.5 Fixed-point parallel implementation of the box-constrained DCD algorithm

We have shown that the serial implementation of the box-constrained DCD algorithm requires very few hardware resources, even for a large number of users. However, due to the sequential processing, this architecture might not provide satisfactory data throughput. In this section, we assume that the elements of vector $\boldsymbol{r}$ are stored in $\boldsymbol{\theta}$ RAM and are updated sequentially. We consider a change in the method of storing elements of the residual vector $\boldsymbol{r}$. These elements are now stored in registers, which allows all K elements of $\boldsymbol{r}$ to be updated in one clock cycle for each "successful" iteration. In addition, all elements in column $\boldsymbol{R}(:,j)$ are accessible simultaneously, which can be achieved in two ways. In the first way, called the ***R**-in-Register*, *the elements of the column* $\boldsymbol{R}(:,j)$ are stored in registers. In the second way, called ***R**-in-RAM*, the elements of column $\boldsymbol{R}(:,j)$ are stored in block RAMs[9].

The box-constrained DCD algorithm in parallel architecture is presented in table 7-6. In state 0, the control signals are initialized. The elements of vector $\boldsymbol{r}$ are stored in the registers, and the elements of matrix $\boldsymbol{R}$ are stored into registers or RAM. In state 1, if $m \neq 0$, the left-shifting operations of the elements of $\boldsymbol{r}$ are executed simultaneously in a single cycle. Meanwhile, the step size d is updated, and the bit counter m is decremented by one. In state 2, the Master State Machine passes the elements of $\boldsymbol{R}(:,j)$ and vector $\boldsymbol{r}$ to the $\boldsymbol{r}$ Update Logic. The time needed to assess the elements and compute $\boldsymbol{r}_t$ is approximately two clock cycles. In state 3, the Master State Machine compares $\boldsymbol{R}(j,j)$ (right-shifted) and $\boldsymbol{r}(j)$. In addition, the $\boldsymbol{x}$ Update Logic reads $x(j)$ from $\boldsymbol{x}$ RAM, pre-updates $x(j)$ and stores the updated element in register x_t. It also checks whether x_t is in the range $[-H,H]$ $(\alpha=0)$ or not $(\alpha=1)$ before proceeding to state 4. State 4 checks c and α to decide whether the iteration is successful. If it is successful, $\boldsymbol{r}$, $x(j)$ and index j are updated. The elements of vector $\boldsymbol{r}$ are updated in the same clock cycle because they are already stored in the registers. In addition, the iteration counter p is incremented by one, and the binary variable Flag is set to 1 to indicate that the current iteration is "successful". The system must then check whether "successful" iteration counter p reaches the maximum number N_u, and if so, the algorithm stops. If not, the system decides whether to update the Flag and which state to proceed to depending on j and Flag.

Table 7-6 Parallel implementation of the box-constrained DCD algorithm

state	Operation	Cycles
0	Initialization: $\boldsymbol{x}=0$, $\boldsymbol{r}=\theta$, $\boldsymbol{m}=M_b$, $p=0, j=1$, Flag$=0$	
1	if $m=0$, algorithm stops else, $m=m-1$, $d=2^m$, $\boldsymbol{r}=2\boldsymbol{r}$	1
2	$\boldsymbol{r}_t=\boldsymbol{r}-\text{sign}(\boldsymbol{r}(j))\cdot\boldsymbol{R}(:,j)$ $c=\boldsymbol{R}(j,j)/2-\lvert\boldsymbol{r}(j)\rvert\times 2^{\Delta m}$ $x_t=x(j)+\text{sign}(\boldsymbol{r}(j))\cdot d$ if $\lvert x_t\rvert\leqslant \boldsymbol{H}$, then $\alpha=0$; else, $\alpha=1$	1
3	$c=R(j,j)/2-\lvert\boldsymbol{r}(j)\rvert$ $x_t=x(j)+\text{sign}(\boldsymbol{r}(j))\cdot d$ if $\lvert x_t\rvert\leqslant H$, then $\alpha=0$; else, $\alpha=1$	1

Table 7–6

state	Operation	Cycles
4	if $c < 0$, and $x(j) = \boldsymbol{x}_t$ $\boldsymbol{r} = \boldsymbol{r}_t$ $p=p+1$, Flag=1 if $p= N_u$, algorithm stops $j = (j)\bmod(K) + 1$ if $j=1$ and Flag=1, then Flag=0, goto state 2 else if $j=1$ and Flag=0, then goto state 1 else, goto state 2	1
Total:	$\leqslant 3KN_u + 3K(M_b - 1) + M_b$	

The number of clock cycles required for each state is shown in table 7–6. Three clock cycles are required in each iteration. The upper bound on the number of the cycles in table 7–6 can be considered as the worst-case complexity of the parallel implementation of the fixed-point box-constrained DCD algorithm. The worst-case situation occurs when only the last bit has N_u successful iterations. This means that the calculations of the first $(M_b - 1)$ bits do not contain any successful iterations and therefore require $(M_b - 1)3K$ cycles. The worst case for calculating the last bit ($m = 1$) occurs when only one successful iteration occurs among the K iterations ($j = 1, \cdots, K$). This case requires $3K$ cycles for the comparison and to pre-update residual vector $\boldsymbol{r}$ and element $x(j)$. In total, N_u successful passes require $3KN_u$ cycles. In addition to the above required cycles, M_b cycles are required to update the step size in the whole process. Therefore, the number of clock cycles of the fixed-point box-constrained DCD algorithm is upper bounded by $3KN_u + 3K(M_b - 1) + M_b$.

***R*-in-Register:** The elements of matrix $\boldsymbol{R}$ and vector $\boldsymbol{r}$ are stored in registers (see figure 7–8). The one-bit left shift of vector $\boldsymbol{r}$ and the update of step size d are performed in state 1. In state 2, the elements in vector $\boldsymbol{r}$ and $\boldsymbol{R}(:,j)$ are read out to update the residual vector $\boldsymbol{r}$. In state 3, the comparison of $r(j)$ and $\boldsymbol{R}(j,j)$ and the pre-update of $x(j)$ are executed in a single clock cycle. In state 4, the final updated elements in vector $\boldsymbol{r}$ and $x(j)$ are obtained in a single clock cycle. In this case, there is a high area expense because all the elements of matrix $\boldsymbol{R}$ are stored in registers. To further improve the design, we consider storing $\boldsymbol{R}$ in a block of RAM.

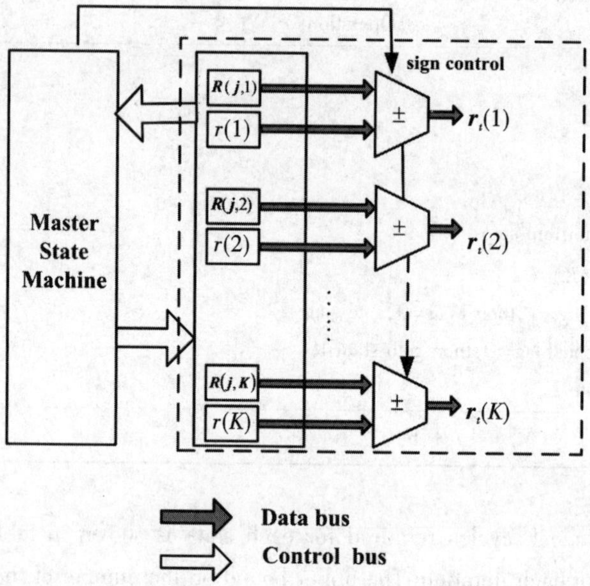

Figure 7–8 ***r*** **Update Logic architecture in *R*–in–Register**

R*-in-RAM**: A row of elements of matrix ***R is stored in an array of RAMs instead of registers to achieve a significant reduction in the required number of slices compared to that required for the case where the elements of ***R*** are stored in registers. This design is illustrated in figure 7–9.

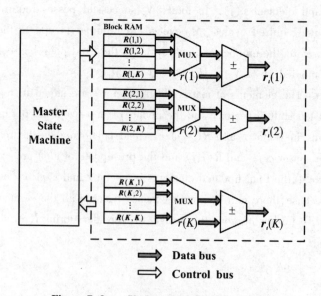

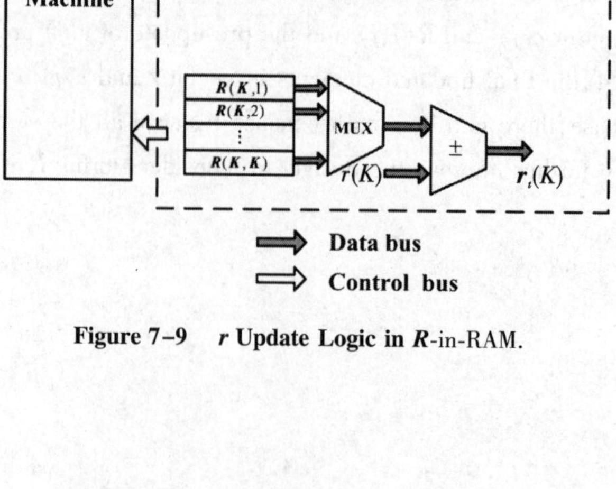

Figure 7–9 ***r*** **Update Logic in *R*-in-RAM.**

The parallel designs have improved throughput compared with that of the serial design of the box-constrained DCD detector in the FPGA implementation. However, the two parallel design implementations are more applicable for scenarios with a small number of users due to the high hardware resources usage. The FPGA resources of both the ***R***-in-Register and ***R***-in-RAM implementations are presented in table 7-7 for the case of K = 16 and M_b = 15. The FPGA resources of the ***R***-in-Register implementation are higher than those of the ***R***-in-RAM implementation because it requires more registers. Therefore, the ***R***-in-Register implementation is suitable for small systems. The ***R***-in-RAM implementation reduces the area usage compared with that of the ***R***-in-Register implementation. However, it still requires more slices than that of the serial architecture box-constrained DCD algorithm implementation.

Table 7-7 FPGA resources of the parallel implementation of the box-constrained DCD algorithm for K= 16 and M_b = 15

FPGA Resources	***R***-in-Register	***R***-in-RAM
Slices	7176	1465
D-FFs	5123	802
LUT4s	5646	2754
Block RAMs	2	18

Exercises

1 Explanation of glossary

1-1 DCD

1-2 state machine

1-3 VHDL

2 Choice question

2-1 The fixed-point DCD algorithm is synthesized and downloaded to this FPGA chip through the Xilinx ISE 8.1i running at the clock frequency ( ).

A. 50 M B. 100 M

C. 200 M D. 400 M

2-2 The FPGA resources of the ***R***-in-Register implementation are ( ) than that of the ***R***-in-RAM implementation.

A. higher B. lower

C. equal D. uncertain

3 Short answer question

3-1 Discuss the advantages of box-constrained DCD-based multiuser detector.

3-2 Simplely describe the five sub-modules in the FPGA architecture of the box-constrained DCD algorithm.

3-3 Analyze the differences between ***R***-in-Register DCD and ***R***-in-RAM DCD.

Reference

[1] VERDU S. Multiuser Detection. 2nd ed. New York, NY, USA: Cambridge University Press, 1998.

[2] BARBERO L G, THOMPSON J S. Rapid prototyping of a fixed-throughput sphere decoder for MIMO Systems. IEEE International Conference on Communications, ICC'06, June. 2006, 7:3082-3087.

[3] HUANG X, LIANG C, MA J. System architecture and implementation of MIMO sphere decoders on FPGA. IEEE Transactions on Very Large Scale Integration (VLSI) systems, Feb. 2008, 16(2):188-197.

[4] CESEAR T, URIBE R. Exploration of least-squares solutions of linear systems of equations with fixed-point arithmetic hardware. In Proceeding of the Software Defined Radio TechnicalConference, Nov. 2005.

[5] ZAKHAROV Y V, TOZER T C. Box-constrained multiuser detection based on multiplication-free coordinate descent optimization. In Proc. Fifth IEEE Workshop on Signal Processing Advances in Wireless Communications, Lisboa, Portugal, Jul. 2004:11-14.

[6] JALDEN J, OTTERSTEN B. An exponential lower bound on the expected complexity of sphere decoding. In Proc. IEEE International Conference on Acoustics, Speech, and Signal Processing (ICASSP), May 2004, 4: iv-393- iv-396.

[7] Xilinx Inc. Virtex-4 Libraries Guide for HDL Designs. UG619(v 14.1), April 24, 2012[2018-08-27]. http://www.xilinx.com/support/documentation/sw_manuals/xilinx14_1/virtex4_hdl.pdf

[8] LIU J, WEAVER B, WHITE G. FPGA implementation of the DCD algorithm. In London Communication Symposium, London, U.K., Sept. 2006:125-128.

[9] LIU J, QUAN Z, ZAKHAROV Y V. Parallel FPGA implementation of DCD algorithm. 15thInternational Conference on Digital Signal Processing, Cardiff, U.K., July. 2007:331-334.

Chapter 8 Box-constrained DCD algorithm for MIMO detection of complex-valued symbols

8.1 Introduction

MIMO communication systems with SM allow the channel capacity to be increased compared to that of SISO communication systems[1]. This requires efficient detection techniques to be used at the receiver. In practical scenarios, we need to consider a reasonable way to achieve the hardware implementation of MIMO detectors, e. g. , using an FPGA platform. The ML MIMO detector provides optimal performance; however, it is complicated for real-time implementation. The ML detector can be directly implemented in hardware for only small MIMO systems with low-order modulation (e. g. , for a 4 transmit and 4 receive antenna MIMO system with QPSK modulation)[2].

The sphere decoder is considered to be a good candidate for hardware implementation of the optimal detector. However, it also becomes more complicated when the size of the system and modulation order increase. In addition, this algorithm requires high complexity at low SNRs. Usually, the decoder's throughput is given at an SNR of 20 dB in [3], [4], whereas at lower SNRs, the throughput is significantly lower. Therefore, sphere decoders are usually implemented on FPGA platforms for small MIMO systems, such as 4×4 systems[4,5]. Moreover, matrix factorization, such as Cholesky decomposition or QR decomposition, and decorrelating detection are required before applying the sphere decoder. This is difficult for real-time hardware implementation[4,6].

The optimal detector and its suboptimal approximations, such as sphere decoders[3,7], provide hard decisions. However, the soft decisions are of interest as they enable further efficient decoding. MIMO detection with soft decision can be based on MMSE detection. Recently, various FPGA designs for MMSE MIMO detection have been reported in [8]-[10]. However, the hardware complexity of a MIMO system increases rapidly with the number of antennas, and only MMSE MIMO detectors with small size were implemented on FPGA platforms. Moreover, the performance of MMSE detectors can be significantly inferior to the optimal detection performance in large size systems.

In this chapter, we consider box-constrained MIMO detection that also provides soft output. The box-constrained detection is well known in application to multiuser CDMA systems[11]. It shows better detection performance than that of MMSE. Moreover, it can be efficiently implemented using DCD iterations[12]. An FPGA design of a box-constrained DCD-based multiuser detector was presented in Chapter 7. However, that design can only be used for real-valued systems, e. g., for systems with BPSK modulation and real-valued system matrix $\boldsymbol{R}$. In MIMO systems, complex-valued modulation schemes, such as QPSK and 16-QAM, are of interest. In this chapter, we present an FPGA design of a complex-valued DCD-based box-constrained MIMO detector of symbols with various QAM modulations and compare it with the designs of MMSE MIMO detectors in terms of the design area, throughput and detection performance.

8.2 System model and box-constrained MIMO detector

We consider an $M_T \times M_R (M_T = M_R)$ MIMO system with M_T transmit and M_R receive antennas in frequency-flat Rayleigh fading channels. The received signal is given by

$$\boldsymbol{y} = \boldsymbol{G}\boldsymbol{x} + \boldsymbol{n} \tag{8.1}$$

where $\boldsymbol{G}$ is an $M_R \times M_T$ channel matrix whose entries are independently and identically distributed (i. i. d.) zero-mean Gaussian random numbers, $\boldsymbol{x}$ is an $M_T \times 1$ vector of transmitted data from a QAM constellation $\mathcal{A}$, and $\boldsymbol{n}$ is the noise vector whose entries are i. i. d. zero-mean Gaussian random numbers. The ML detector solves the quadratic optimization problem with integer constraints:

$$\hat{\boldsymbol{x}}_{\mathrm{ML}} = \underset{x \in \mathcal{A}^{M_T}}{\operatorname{argmin}} \{ \| \boldsymbol{y} - \boldsymbol{G}\boldsymbol{x} \|^2 \} = \underset{x \in \mathcal{A}^{M_T}}{\operatorname{argmin}} \{ \boldsymbol{x}^{\mathrm{H}} \boldsymbol{R} \boldsymbol{x} - 2\boldsymbol{\theta}^{\mathrm{H}} \boldsymbol{x} \} \tag{8.2}$$

where $\boldsymbol{R} = \boldsymbol{G}^{\mathrm{H}}\boldsymbol{G}$ and $\boldsymbol{\theta} = \boldsymbol{G}^{\mathrm{H}}\boldsymbol{y}$. The box-constrained detector relaxes the constraint $\boldsymbol{x} \in \mathcal{A}^{M_T}$ and $\Re\{\boldsymbol{x}\} \in [-H, H]^{M_T}$ and $\Im\{\boldsymbol{x}\} \in [-H, H]^{M_T}$, where H is the maximum magnitude among the real and imaginary elements of the constellation $\mathcal{A}$. For example, for 16-QAM modulation, $H = 3$. Thus, the box-constrained MIMO detector solves the problem

$$\hat{\boldsymbol{x}}_{\mathrm{box}} = \underset{\Re\{x\} \& \Im\{x\} \in [-H, H]^{M_T}}{\operatorname{argmax}} \{ \boldsymbol{x}^{\mathrm{H}} \boldsymbol{R} \boldsymbol{x} - 2\boldsymbol{\theta}^{\mathrm{H}} \boldsymbol{x} \}. \tag{8.3}$$

This solution provides a soft output $\hat{\boldsymbol{x}}_{\mathrm{box}}$, which is then mapped into $\mathcal{A}^{M_T}$. Solving (8.3) is equivalent to solving the normal equations

$$\boldsymbol{R}\boldsymbol{x} = \boldsymbol{\theta} \tag{8.4}$$

with the box constraint in solution $\boldsymbol{x}$.

8.3 FPGA implementation of DCD-based box-constrained MIMO detector

The DCD algorithm solves the normal equations (8.4) using coordinate descent iterations with a varying power-of-two step size[13]. Two types of iterations occur: successful and unsuccessful. In a successful iteration, one element of the solution vector $\boldsymbol{x}$ and all elements of the residual vector $\boldsymbol{r} = \boldsymbol{\theta} - \boldsymbol{R}\boldsymbol{x}$ are updated; these iterations contribute most to the algorithm complexity. In an unsuccessful iteration, there is no update. The maximum number of successful iterations (updates) N_u is predefined. Another parameter M_b, which indicates the number of bits representing elements of $\boldsymbol{x}$, is also predefined and controls the final accuracy of the solution. It also determines how many times the step size d is reduced by a factor of two.

The structure of the box-constrained DCD algorithm for complex-valued modulation schemes is optimized for FPGA implementation and presented in table 8-1. The architecture of the DCD processor is shown in figure 8.1.

Table 8-1 Complex-valued box-constrained DCD algorithm.

state	Operation	Cycles
0	Initialization: $\boldsymbol{x} = \bar{\boldsymbol{x}}, \boldsymbol{r} = \boldsymbol{\theta} - \boldsymbol{R}\boldsymbol{x}$, $m = M_b$, $p=0$, $\Delta m=0, s=1, j=1$, Flag$=0$	
1	if $m=0$, algorithm stops else, $m=m-1, d=2^m$, $\Delta m=\Delta m+1$	1
2	if $s=1$, then $r_t = \Re(r(j))$; else, $r_t = \Im(r(j))$ $c = R(j,j)/2 - \mid r(j) \mid \times 2^{\Delta m}$ $x_t = x(j) + \text{sign}(r_t) \cdot s \cdot d$ if $\mid \Re(x_t) \mid \leqslant H$ and $\mid \Im(x_t) \mid \leqslant H$, then $\alpha = 0$; else $\alpha = 1$	1
3	if $c < 0$, and $\alpha = 0$, goto state 4 else, goto state 5	1
4	$x(j) = x_t$ $\boldsymbol{r} = \boldsymbol{r} \times 2^{\Delta m} - \text{sign}(r_t) s \boldsymbol{R}(:, j)$ $\Delta m=0, p=p+1$, Flag$=1$ if $p = N_u$, algorithm stops	M_T
5	if $s=1$, then $s=i$, goto state 2 else, $s=1$, $j=(j) \bmod (M_T)+1$ if $j=1$ and Flag$=1$, then Flag$=0$ goto state 2 else if $j=1$ and Flag$=0$, then goto state 1 else, goto state 2	1
Total:	$\leqslant 7M_T N_u + 6M_T(M_b - 1) + M_b$ Cycles	

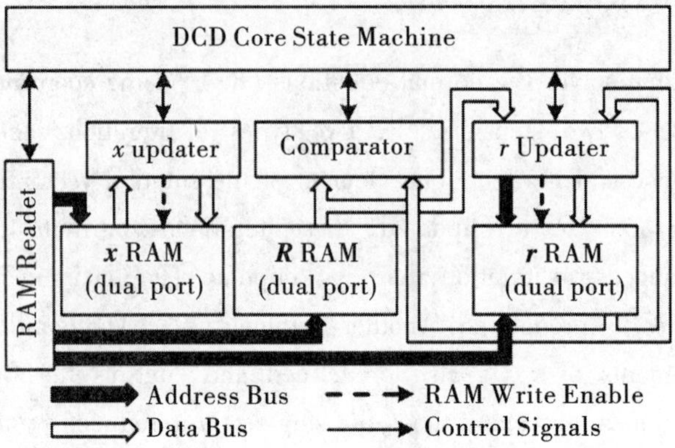

Figure 8-1 Block diagram of the DCD processor

A DCD Core State Machine controls the operations of the algorithm. In state 0, it initializes several control signals, as shown in table 8-1. The residual vector $\boldsymbol{r} = \boldsymbol{\theta} - \boldsymbol{R}\bar{\boldsymbol{x}}$, where $\bar{\boldsymbol{x}}$ is the initialization of $\boldsymbol{x}$. In this chapter, $\bar{\boldsymbol{x}}$ is set to zero, and $\boldsymbol{r}$ is set equal to $\boldsymbol{\theta}$. Bit counter m is set to M_b, successful iteration counter p is set to 0, pre-scaling counter Δm is set to 0, signal s is set to 1 for real components and to i for imaginary components of element $r(j)$, element index j is set to 1, and successful iteration indicator "Flag" is set to 0.

In state 1, bit counter m, step size $d = 2^m$, and pre-scaling counter Δm are updated. The condition $m = 0$ implies that the least significant bits of elements of the solution vector have been decided, and the DCD processor stops. Otherwise, the algorithm proceeds to state 2.

In state 2, RAM Reader asserts the addresses of $r(j)$ in $\boldsymbol{\theta}$ RAM, $R(j,j)$ in $\boldsymbol{R}$ RAM, and $x(j)$ in $\boldsymbol{x}$ RAM. Then, according to s, the Comparator selects the real (if $s = 1$) or imaginary (if $s = i$) component of element $r(j)$. If the vector $\boldsymbol{r}$ has not been prescaled for a new mth bit, the Comparator scales r_t. It then performs the comparison and passes the sign bit of the result c to DCD Core State Machine. Simultaneously, the $\boldsymbol{x}$ Updater reads $x(j)$ from $\boldsymbol{x}$ RAM, pre-updates the (real or imaginary according to s) component of $x(j)$, stores the updated element as x_t in a register, and checks whether it is in the range $[-H, H]$ ($\alpha = 0$) or not ($\alpha = 1$), where the value of H depends on the modulation, i. e., $H = 1$ for M-PSK.

In state 3, DCD Core State Machine examines the comparison results c and α to decide which step to proceed to. If $c < 0$ and $\alpha = 0$ (the iteration is successful), the algorithm proceeds to state 4 to update $x(j)$ and $\boldsymbol{r}$; otherwise, it proceeds to state 5 without updates (the iteration is unsuccessful).

In state 4, $\boldsymbol{r}$ Updater updates $\boldsymbol{r}$. RAM Reader generates the addresses of the elements of column $\boldsymbol{R}(:,j)$ and vector $\boldsymbol{r}$ in $\boldsymbol{R}$ RAM and $\boldsymbol{\theta}$ RAM, respectively. $\boldsymbol{r}$ Updater reads the elements of $\boldsymbol{r}$ from $\boldsymbol{\theta}$ RAM, updates them, and writes the result back. $\boldsymbol{r}$ Updater has two adders to simultaneously update the real and imaginary components of $\boldsymbol{r}$. $\boldsymbol{x}$ Updater writes x_t into $\boldsymbol{x}$ RAM to *replace* $x(j)$. Then, DCD Core State Machine sets the pre-scaling counter Δm to 0 and the variable Flag to 1, indicating a successful iteration. The counter of successful iterations p is also updated. If p is equal to the predefined limit N_u, the DCD processor stops; otherwise, it proceeds to state 5.

In state 5, DCD Core State Machine first tests s to decide which component of $\boldsymbol{x}$ should be analyzed next. Then, it updates the index j and signal s and decides whether to update the Flag and which subsequent state to proceed to depending on j and Flag.

The number of clock cycles required in each state is shown in table 8-1. Pipelining is used in state 4, so the design requires M_T cycles to update $x(j)$ and all elements in $\boldsymbol{r}$. The other states require a single cycle each. The total number of cycles for solving a system of equations varies depending on the system size, the condition number of the system matrix, and the algorithm parameters, N_u and M_b. For a given M_T, N_u and M_b, the number of cycles can be considered to be a random number with an upper bound corresponding to a worst-case scenario. The upper bound can be shown to be $7M_TN_u+6M_T(M_b-1)+M_b$, or for high N_u, approximately $7M_TN_u$. This corresponds to the unlikely situation where in every pass, only one successful iteration (one update) occurs. In a typical situation, there are many successful iterations in every pass, so the average number of clock cycles is smaller, as explained below.

Each element of a complex-valued system of equations is represented using two 16-bit Q15 numbers[14] to represent real and imaginary components; these are limited to the range $[-1,1)$. To avoid overflow, the real and imaginary components of $\boldsymbol{r}$ and $\boldsymbol{x}$ are stored using 32-bit fixed-point Q15 format and are limited to the range $[-2^{16},2^{16})$. To obtain a high update rate, the real and imaginary components of $\boldsymbol{r}$ are processed in parallel. $\boldsymbol{R}$ RAM has 32-bit data width and $\boldsymbol{r}$ RAM has 64-bit data width. This enables them to support both components simultaneously at each read or write operation. $\boldsymbol{x}$ RAM has 32-bit data width since only one component of $x(j)$ is required in each iteration.

Table 8-2 presents the FPGA resources required for the box-constrained DCD-based MIMO detector of complex-valued symbols. The area usage for the box-constrained DCD-based algorithm is very small, i. e., 637 to 667 slices for systems of equations of size $M_T=4$ to $M_T=16$. In all cases, the design occupies less than 5% of the chip area, and it does not use any built-in multipliers. This design has a significantly lower area usage than those of the MMSE detectors designed for 4×4 MIMO systems reported in [8]-[10], which require 8513, 7679, and 9474 slices. In addition, the MMSE detectors presented in [8], [9] use 64 and 58 area-expensive multipliers, respectively, compared to the absence of multipliers in the DCD-based algorithm.

Table 8-2 FPGA resources required for box-constrained DCD-based MIMO detector of complex-valued symbols.

Resources	$M_T = 4$	$M_T = 8$	$M_T = 16$
Slices	637 (4.7%)	658 (4.8%)	667 (4.9%)
D-FFs	305 (1.1%)	318 (1.2%)	329 (1.2%)
LUT4s	1033 (3.8%)	1062 (3.9%)	1084 (4.0%)
Block RAMs	5 (3.7%)	5 (3.7%)	5 (3.7%)

8.4 Numerical results

In the AWGN channel, we present numerical results that allow us to estimate the throughput of the proposed design. Specifically, the convergence speed of the design, in terms of the number of updates and number of clock cycles, is demonstrated for 4×4, 8×8, and 16×16 MIMO systems with 16-QAM modulation.

By solving (8.3), the DCD algorithm ($M_b = 15$) obtains the solution $\hat{\boldsymbol{x}}_{\text{box}}$. The misalignment between estimated data vector $\hat{\boldsymbol{x}}_{\text{box}}$ and transmit data vector $\boldsymbol{x}$ is calculated as

$$\xi = \frac{\| \hat{\boldsymbol{x}}_{\text{box}} - \boldsymbol{x} \|^2}{\| \boldsymbol{x} \|^2} \tag{8.5}$$

The misalignment is averaged over $T = 1000$ simulation trials and is given in decibels by

$$\bar{\xi} = 10\lg\left\{ \frac{1}{T} \sum_{t=1}^{T} \frac{\sum_{j=1}^{M_T} | \hat{x}_{\text{box}}(j) - x(j) |^2}{\sum_{j=1}^{M_T} | x(j) |^2} \right\} \tag{8.6}$$

This misalignment is plotted against the number of updates N_u in figure 8-2, which is obtained from MATLAB.

For comparison, we also show the results for the MMSE MIMO detector, which is implemented using the DCD algorithm. This MMSE detector algorithm is different from the box-constrained detector in that the comparison with the threshold H in states 2 and 3 in table 8-1 is removed and the matrix $\boldsymbol{R}$ is replaced by $\boldsymbol{R} + \frac{1}{\text{SNR}}\boldsymbol{I}$, where $\boldsymbol{I}$ is an $M_T \times M_T$ identity matrix. The box-constrained solution provides significantly lower misalignment than that of the MMSE solution, and the difference in performance increases as the system size M_T increases.

Figure 8-3 shows the misalignment versus the number of cycles, which is obtained from the FPGA platform. By comparing figure 8-3 and figure 8-2 for a fixed misalignment, i. e., −25 dB, we can conclude that one update requires an average of approximately $2.5M_T$, $2M_T$ and

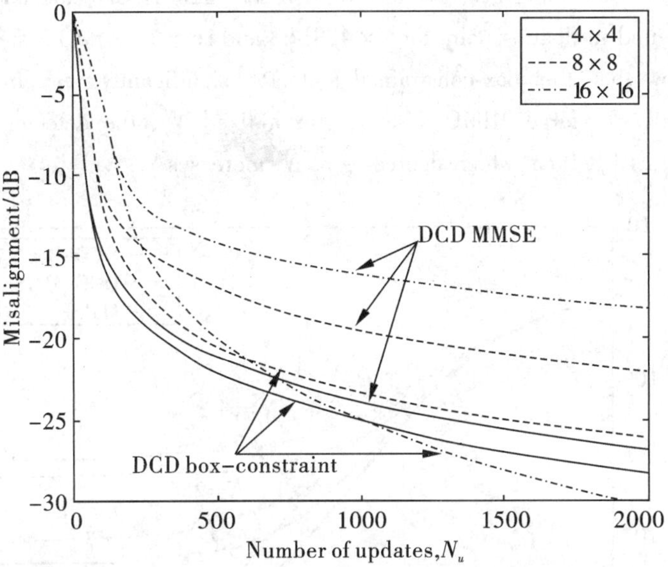

Figure 8-2 Misalignment vs number of updates N_u in 16-QAM MIMO systems

1. $7M_T$ cycles for 4 × 4, 8 × 8 and 16 × 16 MIMO systems, respectively. This requirement is significantly lower than that in the worst-case scenario discussed above. Thus, the total number of clocks required is approximately 2. $5M_TN_u$ for 4 × 4 MIMO systems, $2M_TN_u$ for 8 × 8 MIMO systems, and 1. $7M_TN_u$ for 16 × 16 MIMO systems.

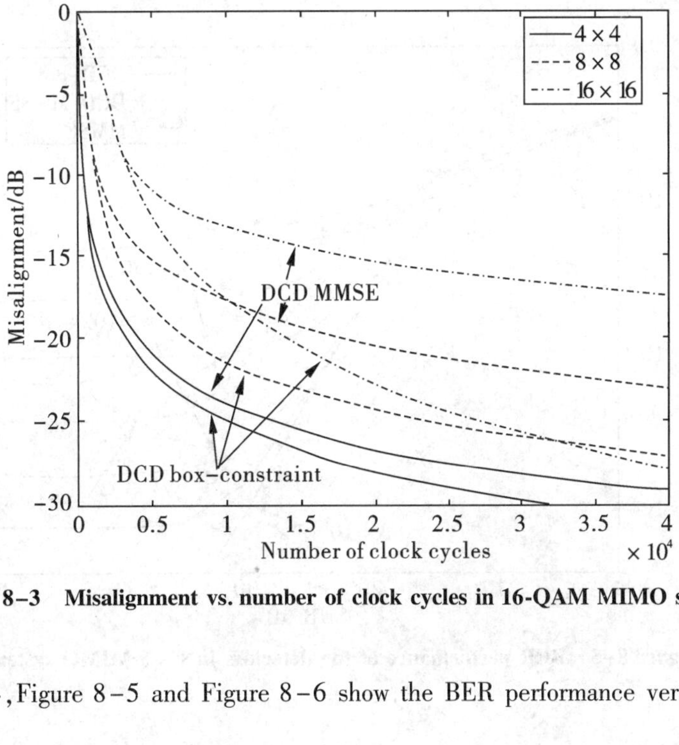

Figure 8-3 Misalignment vs. number of clock cycles in 16-QAM MIMO systems

Figure 8-4, Figure 8-5 and Figure 8-6 show the BER performance versus SNR for the box-constrained detector and those of the DCD-based MMSE detector and classic MMSE detector implemented in floating point for 4×4, 8×8 and 16×16 MIMO systems, respectively. The figures show that the box-constrained detector significantly outperforms the MMSE detectors, especially for large MIMO systems. For a fixed N_u, the detectors based on DCD iterations exhibit a BER floor, which decreases as N_u increases.

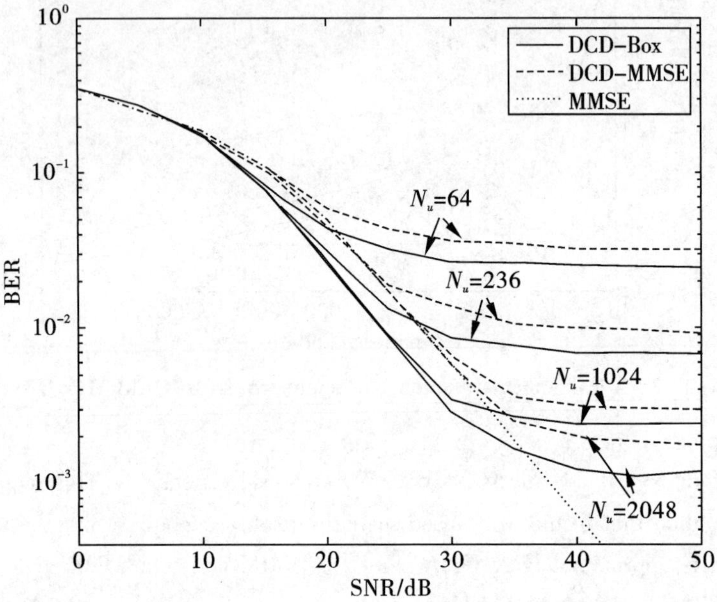

Figure 8-4 BER performance of the detectors in 4×4 MIMO systems

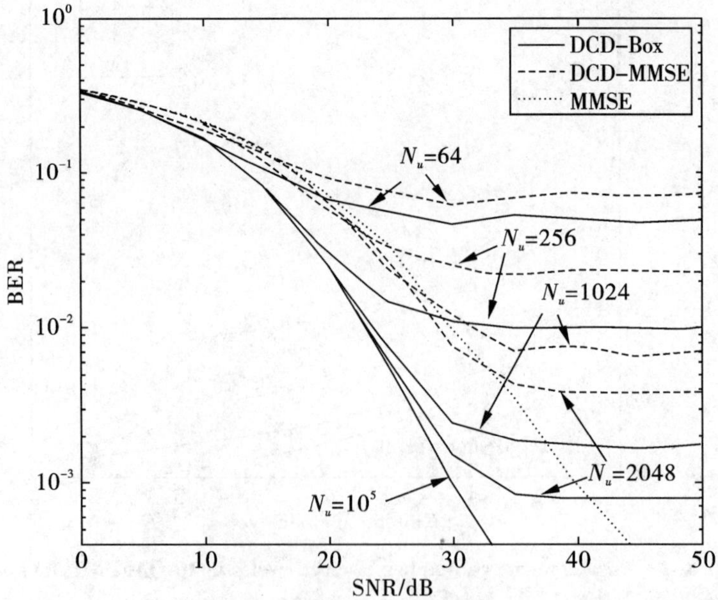

Figure 8-5 BER performance of the detectors in 8×8 MIMO systems

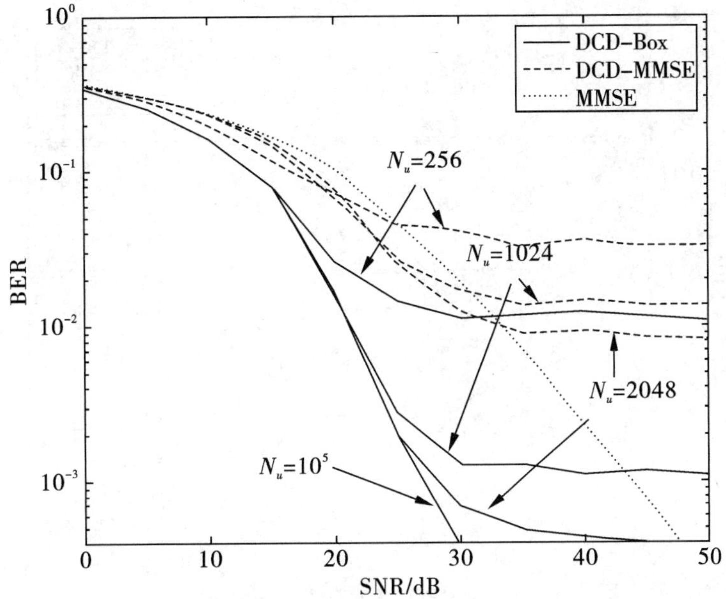

Figure 8-6 BER performance of the detectors in 16×16 **MIMO systems**

Figure 8-7 shows the improvement in BER performance achieved by the box-constrained detector as the number of updates increases for 4×4, 8×8 and 16×16 MIMO systems. In the 4×4 case, figure 8-4 shows that the BER of the MMSE detector is 0.18, 0.05 and 0.006 at SNR = 10 dB, 20 dB and 30 dB, respectively. Figure 8-7 shows that the box-constrained detector can achieve these BERs with approximately N_u = 27, 47 and 410 updates corresponding to 270, 470 and 3800 cycles, respectively. We conclude that, from the perspective of the throughput (the amount of data processed per clock cycle), the proposed design is inferior to the MMSE designs in [8]-[10], which require 270 to 388 cycles. However, due to the very low area usage of our design (less than 5% of the Xilinx Virtex-II Pro Development System[15] with an XC2VP30 (FFT896 package, speed grade 7), we can implement approximately 20 DCD-based box-constrained detectors using the entire area of this FPGA chip. With 20 box-constrained detectors working in parallel, we can reduce the average number of cycles for detection of one MIMO symbol to 14, 24 and 190 cycles at SNR = 10 dB, 20 dB and 30 dB, respectively. These numbers of cycles are significantly lower than those of the MMSE detector. The use of more advanced FPGA chips, such as Virtex-5 XCE5VLX330[16], would increase the number cf parallel DCD-based MIMO detectors to 200. Note that in OFDM MIMO systems, implementation of a number of detectors working in parallel, e.g., one detector per subcarrier, can be beneficial.

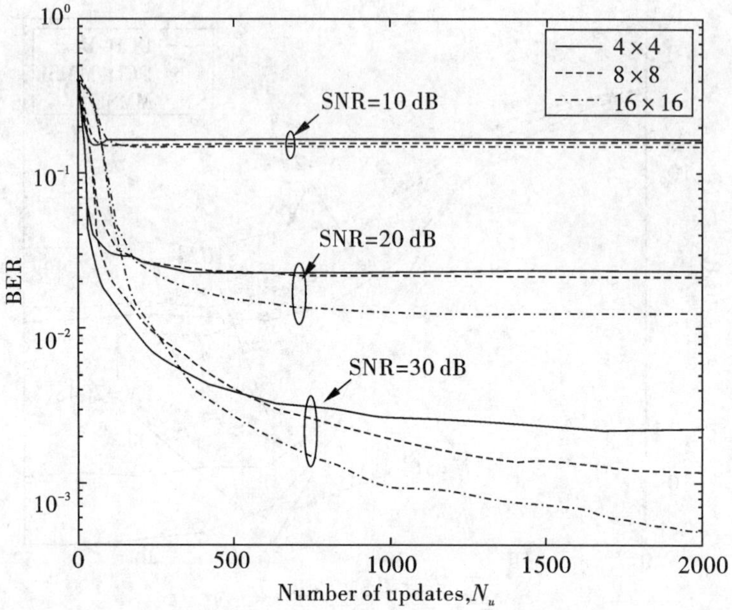

Figure 8–7 BER vs number of updates N_u in 16-QAM MIMO systems

As the size of the MIMO system increases, the proposed design becomes relatively more efficient. In particular, for the 8 × 8 MIMO system, at SNR = 10 dB, 20 dB and 30 dB, the BER performance of the MMSE detector (BER = 0.22, 0.07 and 0.01) is achieved with $N_u = 40$, 68 and 297 updates, i.e., approximately 650, 1100 and 4800 cycles, respectively. For 16 × 16 MIMO systems, the box-constrained detector requires approximately 1800 cycles ($N_u = 68$) at SNR = 10 dB, 2600 cycles ($N_u = 95$) at SNR = 20 dB and 4600 cycles ($N_u = 170$) at SNR = 30 dB to achieve the BER performance of the MMSE detector (BER = 0.25, 0.1 and 0.02, respectively). Table 8–3 shows the cycle count results. The proposed design is shown to be especially useful for large MIMO systems.

Table 8–3 Number of cycles required by the box-constrained DCD-based MIMO detector to achieve the BER performance of the MMSE detector

SNR	4 × 4	8 × 8	16 × 16
10 dB	270	650	1800
20 dB	470	1100	2600
30 dB	3800	4800	4600

Exercises

1 Explanation of glossary

1-1 throughput

1-2 OFDM

1-3 misalignment

2 Choice question

2-1 In the following algorithms, which detector is hard to implement on FPGA platform for large size.

A. Decorrelating B. Maximum Likelihood

C. Decision feedback D. MMSE

2-2 The value of H in the DCD algorithimdepends on the modulation, which is suitable for 16-QAM.

A. 1 B. 3

C. 10 D. 2

3 Short answer question

3-1 Briefly describe the process of FPGA implementation of DCD-based box-constrained MIMO detector.

3-2 Find the fixed-point 2's complement representation with $B = 8$ for the decimal numbers 0. 152 and - 0. 738. Round the binary numbers to 6 bits and compute the corresponding roundoff errors.

3-3 Compare the DCD-MMSE and MMSE in performance and complexity.

Reference

[1] TELATAR I. Capacity of multi-antenna Gaussian channels. European Transactions on Telecommunications, 1999, 10(6):585-595.

[2] KOIKE T, SEKI Y, MURATAH, et al. Prototype implementation of real-time ML detectors for spatial multiplexing transmission. IEICE Transactions on Communications, 2006, 89(3):845-852.

[3] BURG A, BORGMANN M, WENK M, et al. VLSI implementation of MIMO detection using the sphere decoding algorithm. IEEE Journal of Solid-State Circuits, Jul. 2005, 40(7): 1566-1577.

[4] HUANG X, LIANG C, MA J. System architecture and implementation of MIMO sphere

decoders on FPGA. IEEE Transactions on Very Large Scale Integration (VLSI) Systems, 2008, 16(2):188-197.

[5] BARBERO L G, THOMPSON J S. Rapid prototyping of a fixed-throughput sphere decoder for MIMO systems. IEEE Int. Conf. Communications, ICC'2006, 2006, 7:3082-3087.

[6] CESEAR T, URIBE R. Exploration of least-squares solutions of linear systems of equations with fixed-point arithmetic hardware. In Proceeding of the Software Defined Radio Technical Conference, Nov. 2005.

[7] GARRETT D, DAVIS L, TEN BRINKS, et al. Silicon complexity for maximum likelihood MIMO detection using spherical decoding. IEEE J. Solid-State Circuits., Sep. 2004, 39(9):1544-1552.

[8] KIM H S, ZHU W, MOHAMMEDK, et al. An efficient FPGA based MIMO-MMSE detector. EUSIPCO'2007, Poznan, Poland, 2007:1131-1135.

[9] KIM H S, ZHU W, BHATIAJ, et al. A practical, hardware friendly MMSE detector for MIMO-OFDM-based systems. EURASIP Journal and Advances in Signal Processing, 2008: ID 267460, 14 pages.

[10] EILERT J, WU D, LIU D. Implementation of a programmable linear MMSE detector for MIMO-OFDM. IEEE Int. Conf. Acoustis, Speech, and Signal Processing, ICASSP-2008, 2008:5396-5399.

[11] ZAKHAROV Y V, TOZER T C. Box-constrained multiuser detection based on multiplication-free coordinate descent optimization. In Proc. Fifth IEEE Workshop on Signal Processing Advances in Wireless Communications, Lisboa, Portugal, Jul. 2004:11-14.

[12] TAN P H, RASMUSSEN L K, LIM T J. Box-constrained maximum-likelihood detection in CDMA. In Proc. IEEE 51st Vehicular Technology Conf. Spring-2000, Tokyo, Japan, May 2000, 1:517-521.

[13] ZAKHAROV Y V, TOZER T C. Multiplication-free iterative algorithm for LS problem. Eelctronics Letters, Apr. 2004, 40(9):567-569.

[14] XILINX Inc. Virtex-5 FPGA User Guides. UG193 (v3.6) July 27, 2017 [2018-8-27]. http://www.xilinx.com/support/documentation/user_guides/ug193.pdf

[15] XILINX Inc. Virtex-4 Libraries Guide for HDL Designs. 2012 [2018-08-25]. http://www.xilinx.com/support/documentation/sw_manuals/xilinx14_1/virtex4_hdl.pdf.

[16] BATEMAN A, PATERSON-STEPHENS I. The DSP handbook: algorithms, applications and design techniques. New York: Prentice Hall, 2002.

第1章 扩频技术

码分多址(CDMA)系统是通过扩频技术,即把具有一定带宽的传输数据信号用一个大于带宽的高速伪随机码进行调制,使得原传输数据信号的带宽被扩展,再经过载波调制发送出去。本章主要介绍CDMA系统中所用到的基本概念——特征波形和扩频因子,分析扩频序列对频带使用效率的影响以及抗信道衰落特性。

1.1 特征波形

1.1.1 扩频

扩频(SS)通信技术是由一位女演员和一位音乐家在论文中首次提出的。1941年,好莱坞女星赫迪·拉马尔(Hedy Lamarr)和钢琴家乔治·安热尔(George Antheil)提出了一种用于控制鱼雷的安全无线电连接,并获得了美国专利#2.292.387。当时美国军方并没有很重视这项技术,直到20世纪80年代,它作为解决恶劣环境中涉及无线电连接应用的解决方案而逐渐受到欢迎,重新被大家认识。

短距离数据收发器的典型应用包括卫星定位系统(GPS)、3G移动电信、W-LAN(IEEE802.11a,IEEE802.11b,IEE802.11g)和蓝牙。SS技术在平衡通信需求和可用无线电频率资源之间的无休止较量中提供了帮助(因为无线电频谱是有限的,因此是昂贵的资源)。

在Shannon和Hartley信道容量定理中可以清晰看出SS技术的作用:

$$C = B \cdot \log_2(1 + S/N) \tag{1.1}$$

在这个等式中,C是以每秒比特数(bps)为单位的信道容量,它是基于理论比特误码率(BER)的最大数据传输速率。B是所需的以Hz为单位的信道带宽,S/N是信噪功率比值。我们假定代表通信信道允许的信息量C,也表示期望的性能,那么带宽就是要付出的代价,因为频率是有限的资源。S/N比值体现环境条件或物理特性(障碍,干扰的存在,干扰等)。

这一公式较好的适用于恶劣环境(由噪声和干扰引起的低信噪比)中,当信号功率低于噪声功率最低值时,可以通过使用或给予更多的带宽(高B)来保持或者甚至提高通信性能(高C)。

通过把对数基数从2改为e(纳皮尔对数)来对上述等式进行修改,则等式为

$\ln = \log_e$：

$$C/B = (1/\ln 2) \cdot \ln(1 + S/N) = 1.443 \cdot \ln(1 + S/N) \tag{1.2}$$

通过麦克劳林级数展开

$$\ln(1 + x) = x - x^2/2 + x^3/3 - x^4/4 + \cdots + (-1)^{k+1} x^k/k + \cdots \tag{1.3}$$

$$C/B = 1.443 \cdot (S/N - 1/2 \cdot (S/N)^2 + 1/3 \cdot (S/N)^3 - \cdots) \tag{1.4}$$

在扩频应用中 S/N 数值通常较低。(正如之前提到的,信号功率甚至可以低于噪声电平。)假定噪声较大使得信噪比 $S/N \ll 1$,Shannon 的表达式可以化简为:$C/B \approx 1.433 \cdot S/N$。或者大约等于

$$C/B \approx S/N \text{或者} N/S \approx B/C \tag{1.5}$$

因此,在给定信噪比的信道中无差错发送信息,我们只需要提高发射的带宽。这个原理似乎简单明了,但因为基带扩频(扩展到特别大的量级)会使得电子器件作出相应扩频和解扩频操作,具体实现起来非常复杂。

扩频技术在具体实施时有多种方案,但是思路相同:把密钥(也称为码或序列)加入到通信信道。插入码的方式正好定义了之前所讨论的扩频技术。术语“扩频”是指信号带宽被扩展几个数量级,在信道中加入密钥即可实现扩频。

扩频是通过插入更高频率的信号将基带信号扩展到一个更宽的频带内的射频通信系统(图 1-1),即发射信号的能量被扩展到一个到更宽的频带内使其看起来如同噪声一样。扩展带宽和初始信号之间的比值称为处理增益(dB)。典型的扩频处理增益可以从 10 dB 到 60 dB。

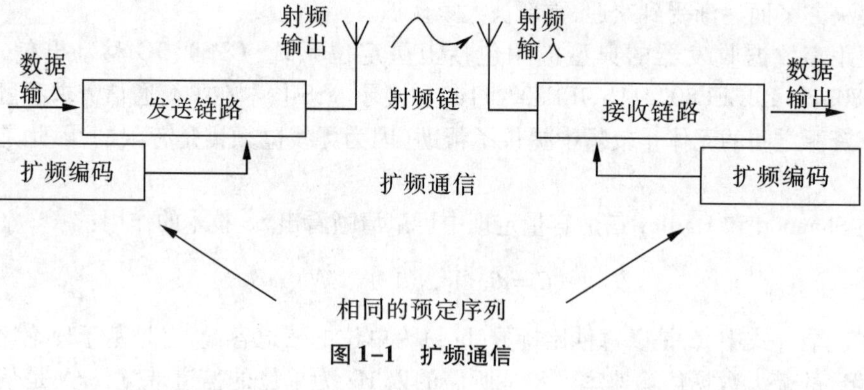

图 1-1　扩频通信

注:采用扩频技术,在天线之前发射链路的某处简单地引入相应的扩频编码,这个过程称为扩频,处理结果将信息扩散到一个更宽的频带内。在接收链路中数据恢复之前移去扩频码,称为解扩频。解扩频是在信号的原始带宽上重新构建信息。显然,在信息传输通路的两端需要预先知道扩频编码(在一些情况下,它应该仅仅被两个当事人知道)。

每个符号的码片数 N 被称为扩频因子、扩频增益或处理增益,有以下作用:

1)对于固定周期的特征波形,其带宽与 N 成正比。

2)对于给定的信噪比,高斯白噪声信道中的单用户误码率与 N 无关。

3)在正交同步直接序列扩频 CDMA 系统中,可支持的用户数小于或等于 N。

4)较大 N 值有助于保护系统隐私,因为它们可以阻止非期望接收者揭示特征波形来窃听所传输的信息。

1.1.2　扩频操作后的带宽效果

图 1-2 描绘了信号带宽在扩频调制后的变化。

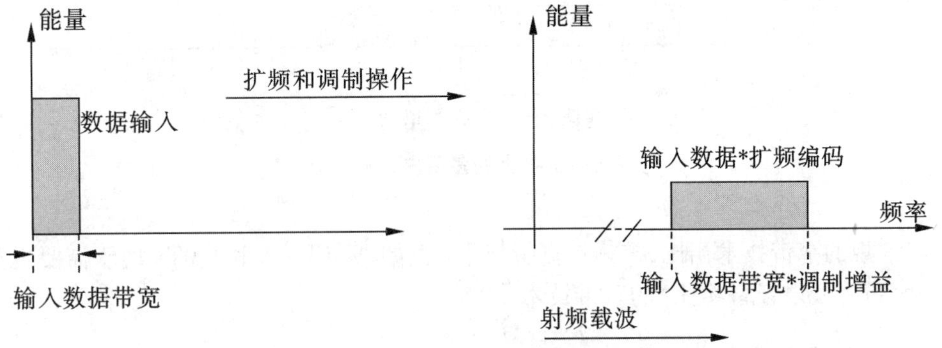

图 1-2　扩频和调制操作

扩频调制作用于通用调制器(如 BPSK)的前端或直接转换。没有收到扩频编码的信号将保持不变,不扩频。

1.1.3　解扩频操作后的带宽效果

同样的,解扩频过程如图 1-3 所示。

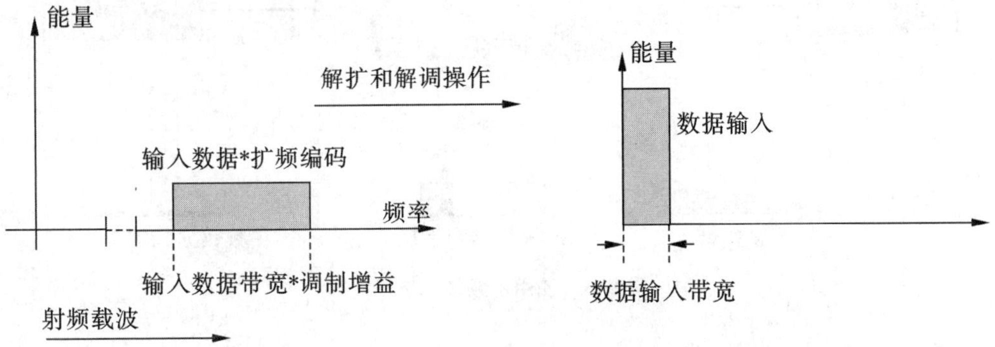

图 1-3　解扩和解调操作

解扩频通常在解调之前进行。在传输过程中加入的信号(如干扰或阻塞)将在解扩处理中被扩频。

1.1.4 多个用户同时共享带宽可以弥补扩频造成的带宽浪费

由于扩频占用更宽的频带(通过前面提到的“处理增益”因子),浪费了有限的频率资源。然而,过度使用的频带资源可以通过多个用户共享同一扩大了的频带得到补偿(图1-4)。

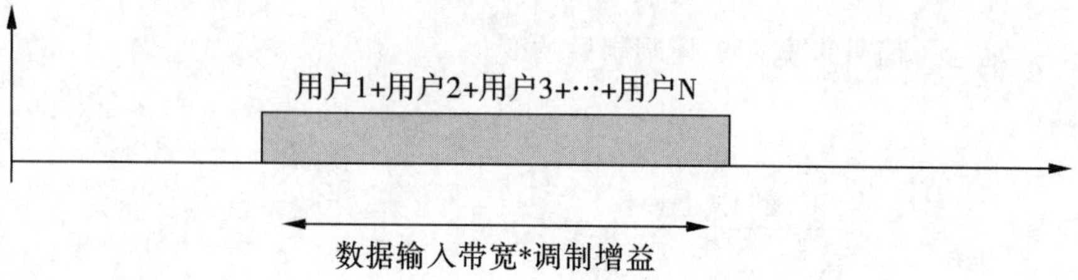

图1-4 扩大的多用户共享频带

与常规的窄带技术相比,扩频是宽带技术。例如,W-CDMA 和 UMTS 属于需要更宽频带(与窄带无线电设备相比)的宽带技术。

1.1.5 抗干扰和抗阻塞效果

获得较高的抗干扰和抗阻塞特性是扩频的优势。因为干扰和阻塞信号不带有扩频因子,所以被抑制掉。解扩频处理之后只有包含扩频因子的期望信号出现在接收器内。

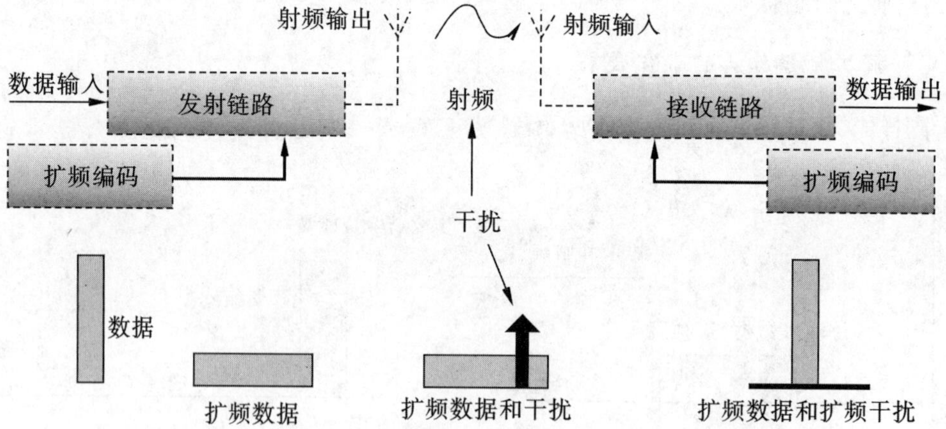

图1-5 使用扩频技术抗干扰

干扰信号可以是窄带,也可是是宽带的;如果干扰信号不包含扩频因子,解扩频之后可以忽略其影响。这种抑制能力同样适用于不具有正确扩频因子的扩频信号,正因为这一点,扩频通信允许不同用户共享同一频带(例如 CDMA)。请注意,扩频是一种宽带技术,但反过来并不成立。也就是说,宽带技术并非都是扩频技术。

1.1.6 拦截抑制

拦截抑制是扩频技术提供的第二个优势(图1-6)。因为没有授权的用户不知道扩频原始信号的扩频因子,因此无法对其进行解码。没有正确的扩频因子,扩频信号将呈现为噪声或干扰(如果扩频因子很短,可以通过扫描方法破解)。另外,扩频通信允许信号低于噪声基底,因为扩频处理降低了频谱密度(总能量相同,但是被扩展到整个频带内)。这样,可以将信息隐藏起来,这是直接序列扩频的显著特点。其他接收器无法“看到”传送信息,只能检测到噪声电平有略微增加!

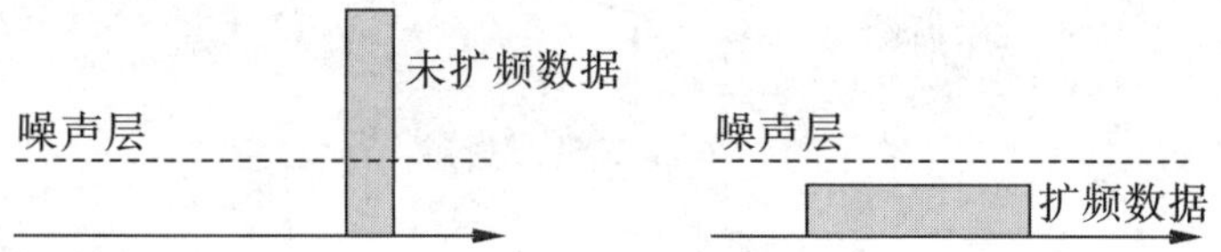

图1-6 拦截抑制

1.1.7 衰落抑制(多径效应)

无线信道通常是多径传播,即信号从发射端到接收端有不止一条路径。这些路径是由于空气反射或折射以及从地面或建筑物等物体的反射产生的(图1-7)。

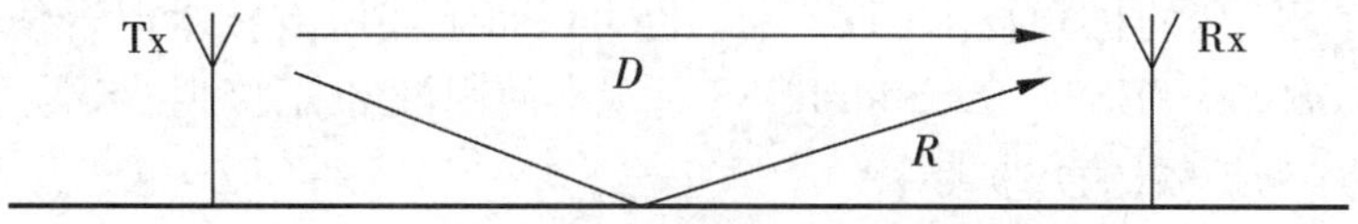

图1-7 衰落抑制

反射路径(R)对直达路径(D)产生干扰被称为衰落现象。因为解扩频过程与信号D同步,即使信号R包含相同的扩频因子也会被抑制掉。有方法可以解扩频反射路径上的信号,并把解扩频后的信息叠加到主信号上。

1.1.8 扩频和编(解)码

扩频的主要特征是发射机和接收机已知扩频编码。在现代通信中,扩频编码是较长且随机生成的显示“像噪声”的数字序列。它们可以重新生成,否则接收方将无法提取已发送的消息。“几乎随机的”的代码被称为伪随机数(PRN)或序列。最常用的生成伪随机码的方法是反馈移位寄存器。

为了保证有效的扩频通信,PRN序列必须遵守一些规则,如长度,自相关,互相关,正交性和位平衡。常见的PRN序列有Barker,M-Sequence,Gold,Hadamard-Walsh等。注意,复杂的序列集合可以提供更稳定的扩频链接。但是,扩频操作需要更多在速度和性能上更有优势的电子设备。纯数字解扩芯片包含超过几百万个等效的两输入与非门,开关频率是几十兆赫兹。

1.1.9 扩频技术的不同调制方式

通过伪 PRN 插入信道的位置不同得到以下几种扩频调制方式,参考 RF 前端原理图 1-8 作简单介绍。

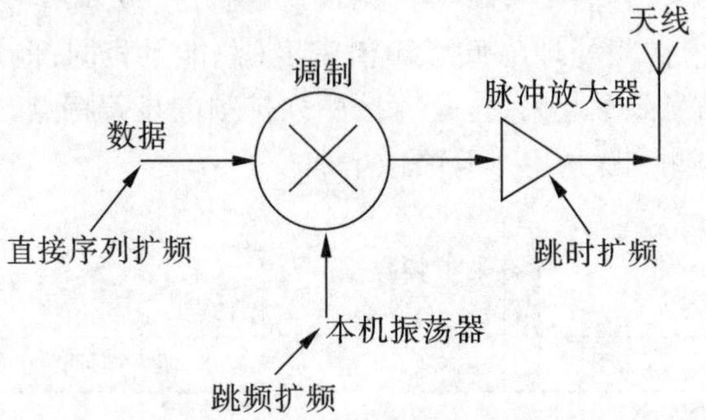

图 1-8　不同位置的不同扩频技术

如果在数据里加入伪随机序列码,则可以得到直接扩频序列(DSSS)。(实际应用中,伪随机序列与信息信号相乘,产生完全被伪随机码"打乱"的数据)。如果伪随机码作用于载波频率上,则我们得到跳频扩频(FHSS)。如果伪随机码作用于本振端,FHSS 伪随机码强制载波按照伪随机序列改变或跳变。如果伪随机序列控制发射信号的开/关,则可以得到时间跳变的扩频技术(THSS)。也可以将上述所有技术混合使用,形成混合扩频技术,如 DSSS + FHSS。DSSS 和 FHSS 是现在使用最多的两种技术。

(1)直接序列扩频(直序扩频)

在这种技术中,伪随机码直接加在载波调制器中的数据上。因此,调制器似乎具有更大的比特率,这是伪随机序列的码片速率。用这样一个码序列调制射频载波的结果是产生一个中心在载波频率、频谱是$[\sin(x)/x]^2$的直接序列调制扩展频谱。这个频谱的主瓣带宽是扩频编码时钟频率的两倍,而旁瓣的带宽等于扩频编码的时钟频率。

图 1-9 是符号速率为 2 bps 的二进制信号的例子。

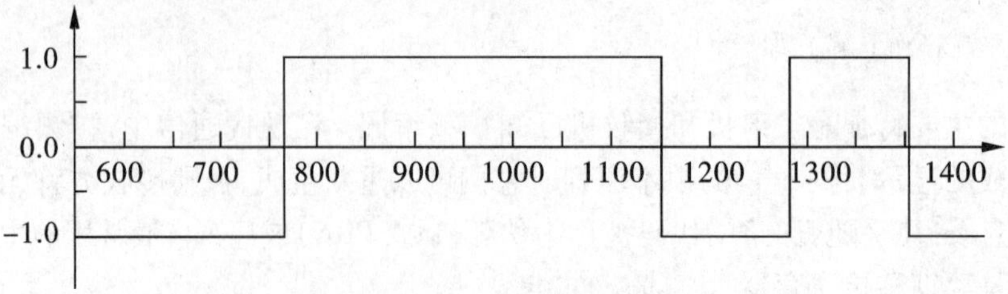

图 1-9　一个二进制信号

对信号进行调制,我们可以把这个序列和一个正弦波相乘,其频谱将如图 1-10 所示。其频谱的主瓣是 2 Hz 宽。符号率越大,信号带宽越大。

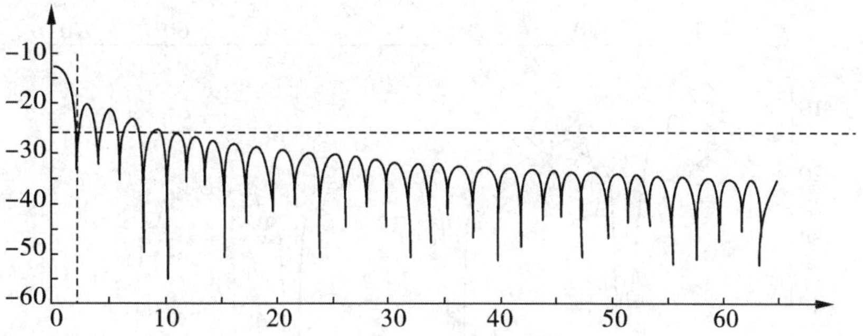

图 1-10 比特率为 2 bps 的二进制信号频谱

二进制序列(图 1-11)的速率是图 1-9 中序列的 8 倍。

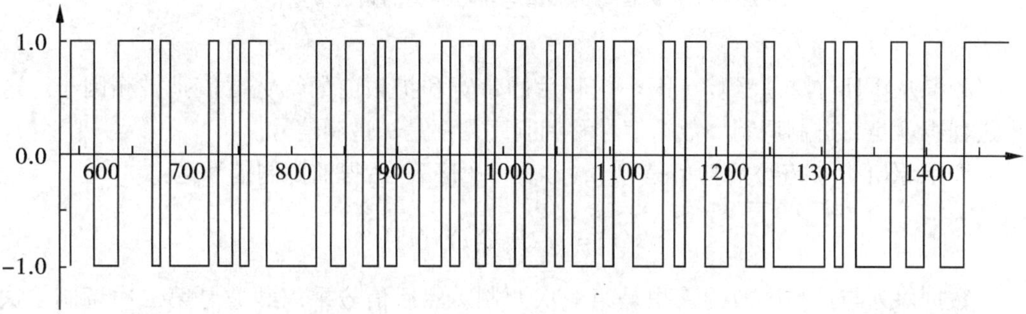

图 1-11 一个用于调制信号序列的二进制新序列

如果我们不用正弦波进行调制,可以用二进制序列(图 1-11)来调制序列 1。调制后的信号如图 1-12 所示。

因为现在比特率更大,这个序列的频谱将有更大的主瓣。

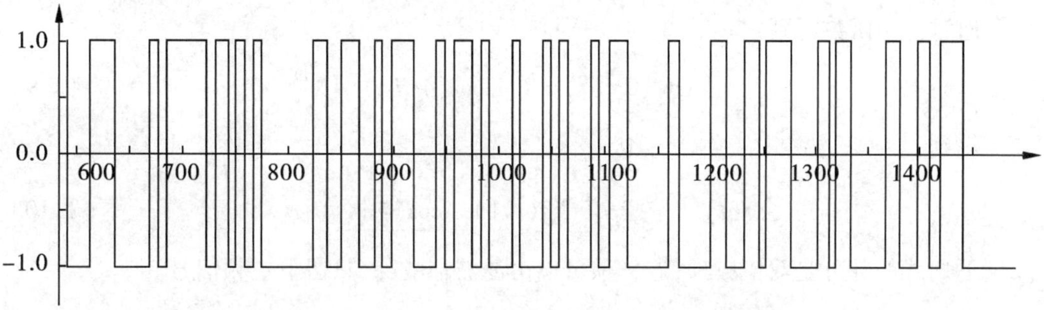

图 1-12 序列 1 中的每一个比特由编码序列替换

信号频谱现在在更大的带宽扩展。扩频后的信号主瓣带宽是 16 赫兹,而不是扩频之前的 2 赫兹。信息序列与扩频序列相乘使得信息序列具有和扩频序列一样的频谱(图 1-13)。

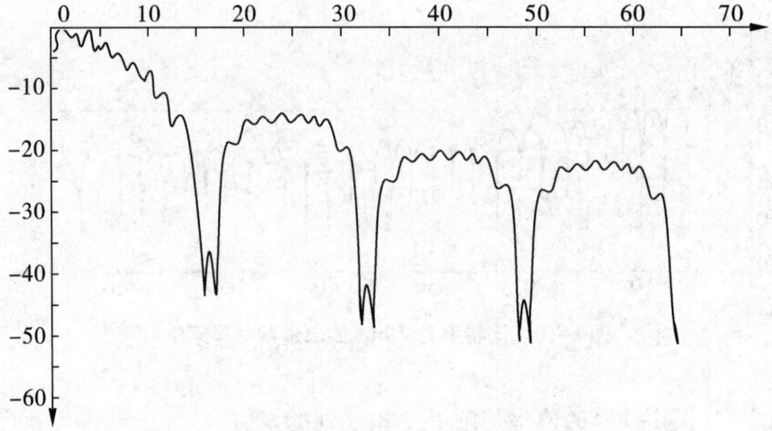

图 1–13　扩频信号的频谱与扩频序列的频谱带宽相同

频谱从 2 Hz 被扩展到 16 Hz。扩频后的带宽和扩频前的带宽比值是扩频因子。这个过程也就是二进制调制。数据信号和调制序列都是二进制信号。

如果原始信号 $d(t)$ 的功率是 P_s，$g(t)$ 表示码序列，所得的调制信号是：

$$s(t)=\sqrt{2P_s}\,d(t)g(t) \tag{1.6}$$

数据序列与扩频序列的乘积是第一次调制。然后信号乘以载波是第二次调制。这里的载体是模拟的信号。

$$s(t)=\sqrt{2P_s}\,d(t)g(t)\sin(2\pi f_c t) \tag{1.7}$$

在接收端，接收信号与载波相乘。可以得到：

$$r(t)=\sqrt{2P_s}\,d(t)g(t)\sin^2(2\pi f_c t) \tag{1.8}$$

通过三角函数得到：

$$\sin^2(2\pi f_c t)=1-\cos(4\pi f_c t) \tag{1.9}$$

得到：

$$r(t)=\sqrt{2P_s}\,d(t)g(t)[1-\underline{\cos(4\pi f_c t)}] \tag{1.10}$$

下划线的部分是双倍频项，这一项可以由滤波器滤除，只留下主频信号。

$$r(t)=\sqrt{2P_s}\,d(t)g(t) \tag{1.11}$$

主频信号与码序列 $g(t)$ 相乘，可以得到

$$r(t)=\sqrt{2P_s}\,d(t)g(t)g(t) \tag{1.12}$$

$g(t)$ 的自相关是一个可以去除的标量，从而得到原始信号。

$$r(t)=\sqrt{2P_s}\,d(t) \tag{1.13}$$

图 1-14 所示是直接序列调制扩频信号的最常见类型。因为直接序列频谱所使用的实际载波和数据调制情况不同，它在频谱形状上会有所不同。图 1-15 是二进制相移键控（BPSK）信号，它是直接序列系统中最常见的调制类型。

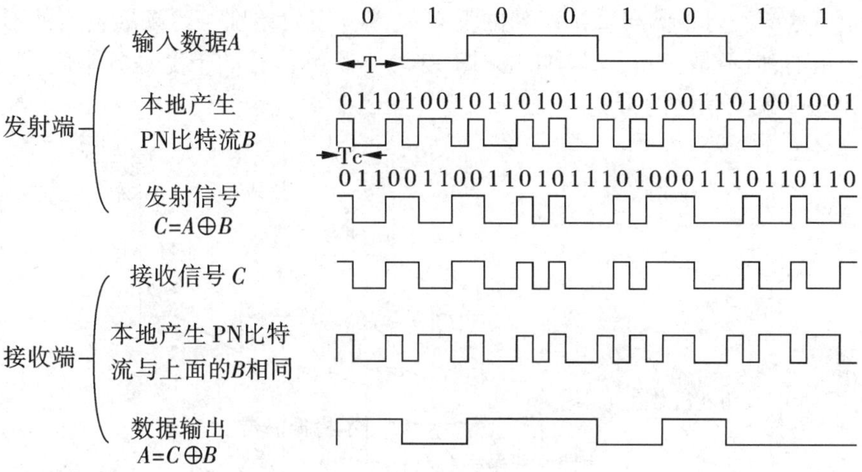

图 1-14　直接序列扩频的示例

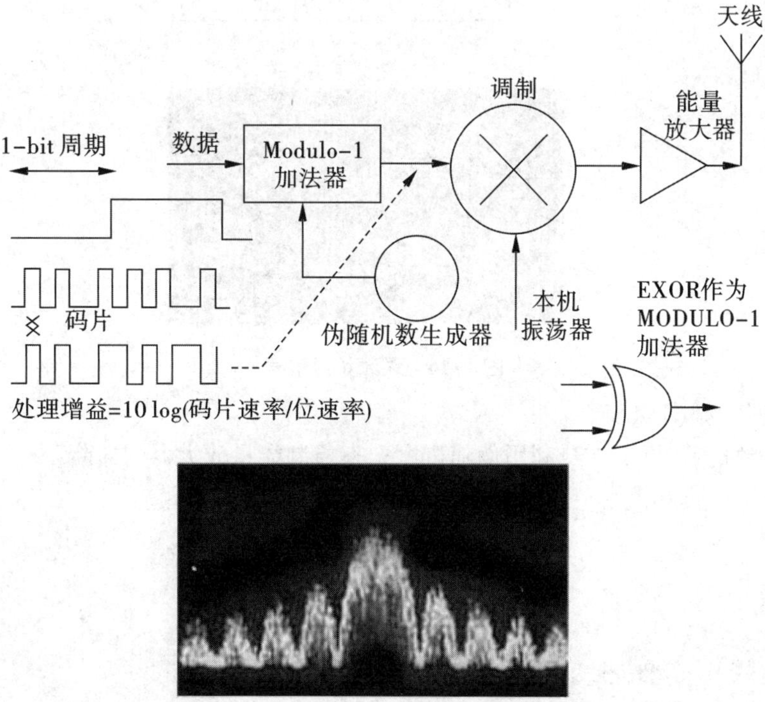

图 1-15　直接序列扩频信号的频谱分析图

（注意原始信号（非扩频）只占用主瓣的一半）

(2)跳频扩频

顾名思义,跳频扩频是载波在一个很宽的频带上按照伪随机码的定义从一个频率跳变到另外一个频率。跳频速度取决于原始信息的数据速率,可分为快速跳频(FFHSS)和慢速调频(LFHSS)。慢速调频(最常见)允许几个连续的数据位调制在相同的频率。另一方面,FFHSS 是在每个数据位内多次调频。

跳频信号的发射频谱与直接序列的发射频谱有很大差别。跳频的包络波形在整个频带是平坦的,而不是$[\sin(x)/x]^2$(见图 1-16)。跳频信号的带宽是频率间隙的 N 倍,其中 N 是每个跳频信道的带宽。

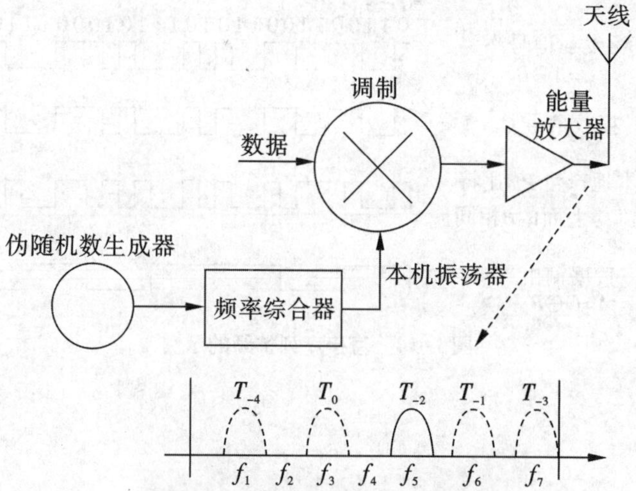

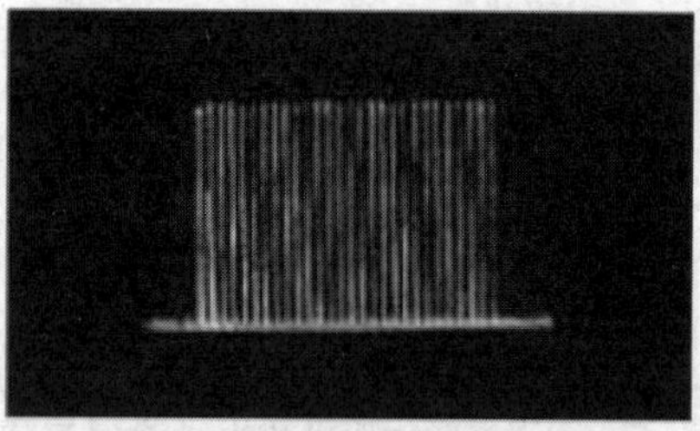

图 1-16　跳频扩频信号

(3)跳时扩频(THSS)

跳时扩频技术(图 1-17)利用伪随机序列来控制能量放大器的通断。

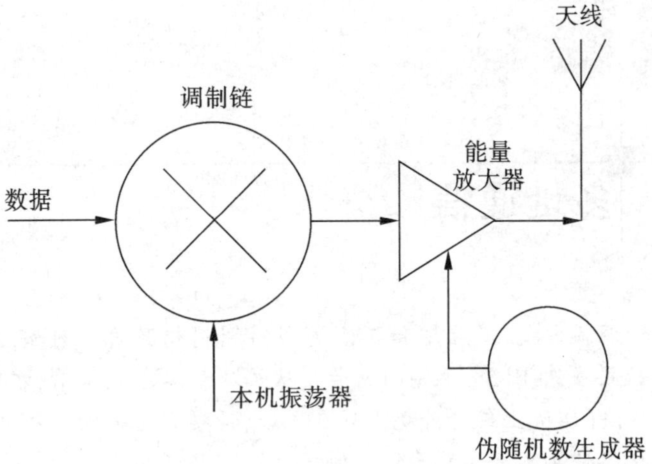

图 1-17　跳时扩频信号

1.2　实现方法

一个完整的扩频通信链路需要涉及各种先进的技术和学科：如射频天线，大功率、高效率功放，低噪声、高度线性低噪声放大器，紧凑型收发器，高分辨率模数转换器（ADC）和数模转换器（DAC），快速低功耗数字信号处理器（DSP）等。虽然设计者和制造商相互竞争，但他们也共同努力促进扩频技术发展。最难以实现的电路是接收机通道，尤其是对直接扩频序列（DSSS）的解扩频部分，因为接收端必须能够重新恢复原始信息并做到实时同步。码的识别也称为相关运算，它是以数字域实现的，需要快速的，高度并行的二进制加法和乘法。到目前为止，接收机设计中最复杂的问题是同步问题。与扩频通信的其他方面相比，开发和改进同步技术投入了更多的时间、精力和资金。

目前能够解决同步问题的方法有许多种，其中很多需要大量独立的元件来实现。DSP 和专用集成电路（ASIC）在这一方面已经取得了最大的突破。DSP 提供高速数学功能，可以分别对扩频信号进行分析，同步和去相关运算。ASIC 芯片通过超大规模集成电路（VLSI）技术降低系统成本，并通过创建基本模块架构使其适用于多种应用。

第 2 章 多址通信

扩频不是一种调制方式,不应该和其他类型的调制相混淆。例如,我们能够利用扩频技术发射一个 FSK 或者 BPSK 调制的信号。从编码基本理论来看,扩频也能作为实现多址通信的一种方法(实际上或者外观上存在多址,链接到同一物理层通信)。多址方式允许多个移动用户同时共享有限的无线频谱。

2.1 多址接入信道

多个发射机在一个通信信道同时发送信息的想法起源于托马斯. 爱迪生 1873 发明的双工式发电机。这个发电机可以在同一方向同一根电缆线同时传输两条电报信息,其中一条信息由极性编码器编码生成,另外一条信息由绝对值编码器编码生成。

目前有很多多个发射机使用同一个信道的多址接入通信的例子,例如多个移动电话发送信号到同一个基站,多个地面站与一个卫星进行通信,多抽头总线,局域网,无线分组网络,交互式有线电视网络等。这些通信方式的共同点是一个接收端会收到多个发射机信号的叠加(图 2-1)。

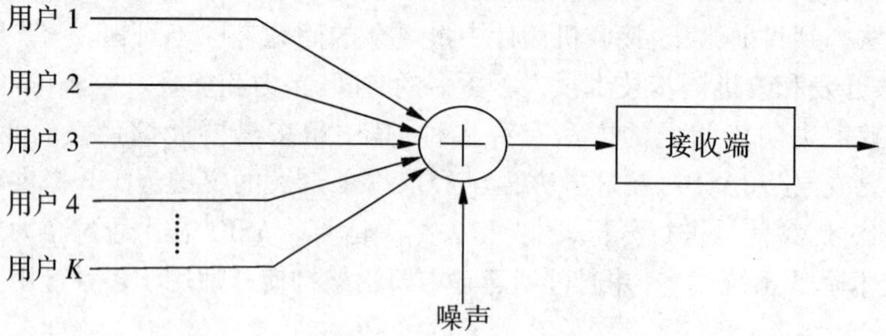

图 2-1　多址接入通信

通常,不同发射机发送的信号的叠加是由于非理想情况而无意产生的。例如电话串扰,以及在蜂窝电话、无线电/电视广播和无线局域网中多个远程发射机同时使用相同无线电频带。通常情况多路复用和多址接入的说法可相互替换,但是多址接入通常是指消息源并不是自主地被操作。多址接入信道中的消息源通常被称为用户。

图 2-1 描绘的是多个用户发送信息给一个接收端的多址接入通信系统,这个系统的接收端也可以是多个接收机,每个接收机对应一个用户发送的信息。多址通信也称为多

点对点通信。同样的信息被发送给多个接收者,例如广播和电视广播或有线电视是把相同信息传递给所有接收者就是这种通信方式。另外一种情况是信息被独立地发送给不同的接收者,例如基站发送信息到移动单元,这种情况就适用于多址接入信道模型,接收机仅对基站发送的众多信息源中的一条感兴趣。

2.2　频分多址和时分多址

20世纪初出现了射频调制,几个无线电通信发射台采用不同频率的载波通信,这样每个发射台发送的信号可以在时间和空间上共存且相互不干扰。之后长途电话也采用了射频调制的方式传输信息。频分复用或频分多址(FDMA)给每个用户分配一个特定的载波频率,每个用户占用的频带不重叠(图2-2)。带通滤波器可以解调每个通信信道。

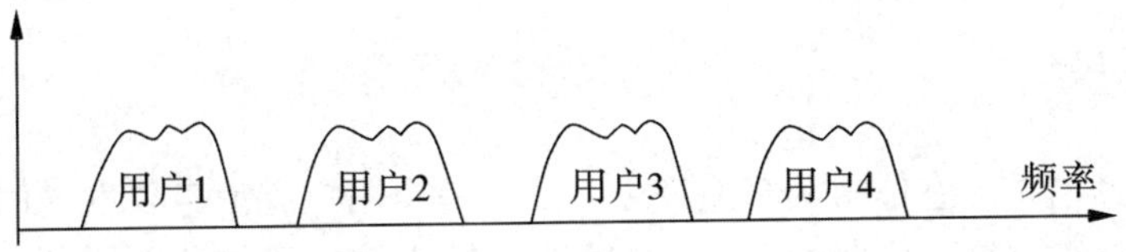

图2-2　频分多址

在时分复用中,不同用户之间被分配一个时隙进行通信(图2-3)。解复用就是接收端在特定时隙转向接收信息。时分的方式不仅适用于多路信息复用,也适用于不同地址用户同时传输信息,这种方式称为时分多址(TDMA)。注意,FDMA可以允许信息在时域内不协同传输:也就是说不需要用户间的时间同步。而TDMA调制要求所有发射机和接收机必须同步。

图2-3　时隙之间具有保护时间的时分多址

从概念上来说,频分和时分多址技术的重要优点是每个用户在独立的且不相互干扰的信道上工作。在数字通信中,多址技术要求每个用户发送的信号相互正交。在非理想情况下,信道或接收机需要在TDMA(图2-3)中加入保护时间,在FDMA(图2-2)中加入频谱保护带从而避免同信道干扰。

为什么要采用多址接入技术而不是把信道分割成相互不干扰的小信道?因为在某个时间段,当用户数量远远大于实际活跃用户的数量时,互不干扰的小信道通信方式必然会浪费信道资源。例如,每个用户被分配一个特定频率的信道,那么在任意时间它只能使用被分配的频段。类似的情况,比如在TDMA中,时隙在无数据传输时处于闲置。

如何动态地为用户分配信道资源?一种方式是建立一个单独的预留信道,如果有用户想要使用该信道就通知接收机,然后接收机仅在活动用户中使用 TDMA 或 FDMA,这种方式会增加运算复杂度,因为它需要预先设定一个反馈信道,通知每个用户在哪个时隙或哪个频段可以发送信息。但是,注意,预留信道是一个多址接入信道,仍然需要处理和以前一样的问题,即如何动态分配信道资源。

FDMA 是无绳电话等应用的理想选择方式。因为 FDMA 的信号处理操作方式简单可以降低无绳电话成本。

FDMA 缺乏内在的多样性,因此蜂窝通信系统使用 FDMA 会使得基站之间产生严重的小区内频率干扰,如果基站间不能严格同步,则基站切换的难度会更大。

对于个人通信系统而言,选择哪种通信方式需要视情况而定。如果预设的个人通信系统更接近无线专业分组交换机而不是蜂窝系统,那么 FDMA 可能是合适的选择。另外采用合并 TDMA 和 FDMA 或 CDMA 系统可以弥补 FDMA 的缺陷。

2.3 随机多址

随机多址通信是动态信道共享途径之一。当一个用户有消息要发送时,它将正常传送,就好像它是该频带内唯一的用户。如果没有其他用户同时发送信息,则消息成功接收。然而,如果用户间不协作通信,就会存在一个信号(在时间和频率上)干扰另一个信号。如果随机多址通信中接收机不能解调同时到达的消息,就通知发送机信息碰撞,并要求重新发送信息。如果发送机接到碰撞信息通知就立即重新发送信息,则碰撞会再次发生。为了解决这个问题,用户可以等待一段时间,然后再次发送消息。目前随机通信系统间的主要区别在于每个碰撞信息重新发送的时延算法。

第一个随机多路访问通信系统是 1969 年为无线电频道设计的 Aloha 系统。以太网在同轴电缆局域网使用 Aloha 另外一种形式,叫做载波侦听多路访问,也就是用户在发送信息前先探听信道使用情况,以免与正在传输的信息相冲突。一般来说,随机多路访问通信比较适合不会有多于一个用户同时传输信息的信道传输。随机多路访问通信的主要理论进展发生在 20 世纪 70 年代到 80 年代中期。轮询是避免同时传输的另一种多路访问策略,也就是接收机会询问信道上的每个发射机是否有传输信息。

2.4 码分多址

到目前为止我们讨论的共享信道方法都是以每个信息有一个特定的时隙或频带为前提的。在随机访问通信中如果违反了这个前提,接收机就不能恢复发生包冲突的传输消息。另外,一个信道里的信息只有正交才不会产生相互干扰。在时间和频率上重叠的信号可以通过正交的方式来传输信息。如图 2-4 和图 2-5 所示,x_1 和 x_2 在时间和频率上重叠(图 2-5)。它们相互正交,内积是零:

$$< x_1, x_2 > = \int_0^T x_1(t) x_2(t) \mathrm{d}t \tag{2.1}$$

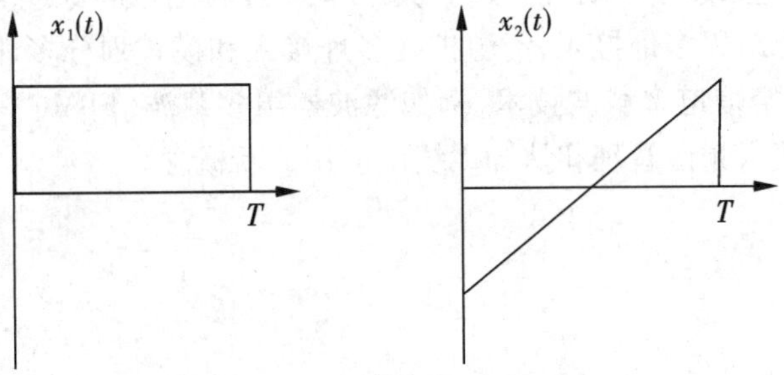

图 2-4　正交信号分配给两个用户

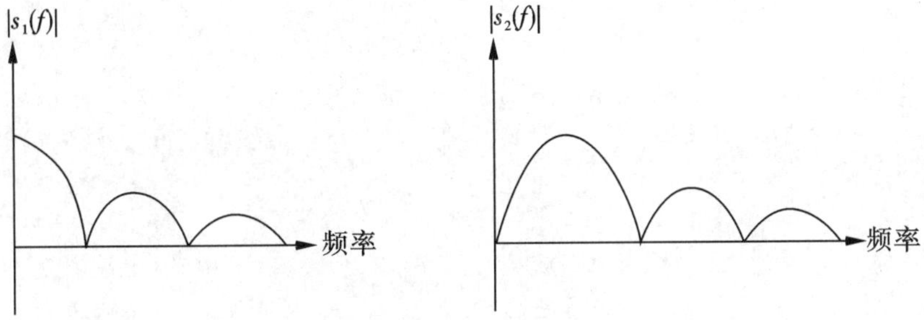

图 2-5　图 2-4 信号波形的傅里叶变换（幅度 $|s_1(f)|>0$，$|s_2(f)|>0$）

接收机收到的通常是有加性背景噪音的两个信号之和。匹配滤波器把接收信号在每个比特周期内分别与 $x_1(t)$ 和 $x_2(t)$ 相关，之后分别把相关器输出与零阈值进行比较。相关器输出会受到背景噪声的影响。然而，基于正交性条件(2.1)和假定的同步性，无论两个信号的相对强度如何，用户 1 相关器的输出完全不受用户 2 的影响。也就是说即使两个用户发送的信号在时间和频率上是重叠的，它的误码率和在两个独立信道中分别发送信息的误码率相同。

以上是码分多址（CDMA）的一个简单例子。所有用户或基站在同一个频段工作。每个用户被分配一个唯一的伪随机码，使得用户在指定的频带上传输信息。正交 CDMA 的特征波形在时域上不重叠，那么它就对应数字 TDMA。数字 TDMA 和 FDMA 系统可以看作是正交 CDMA 的特殊情况，即扩频后的特征波形分别在时域和频域上不重叠。在接收端，不同用户的接收信号通过接收信号与用户扩频后的特征波形互相关来分离。精心设计的互相关性小的扩频序列可以减少同频干扰，CDMA 中被称作多址干扰（MAI），从而达到多用户检测的目标。

2.5　用于 5G 的多址接入通信技术

和高级 LTE 网络相比，第五代（5G）移动蜂窝网络需要更多的网络吞吐量和更高的小区边缘数据速率，高能效和低等待时间。因此在物理层（PHY）无线电空中接口和无线电接入网络上的创新技术是非常重要的。非正交多址作为一种新型无线接入技术引起了学术界和工业界越来越多的研究兴趣。其中中兴通讯、华为、高通、DTmobile 等提出了多用户共享接入、稀疏码多址接入、资源扩展多址接入和模式划分多址接入。与此同时，把频带分割成多个窄信道多载波技术，例如滤波器组多载波(FBMC)和广义频分复用等技术，已成为动态接入频谱管理和认知无线电应用的新概念。

第3章 信道模型

本章介绍信道模型基本概念:路径损耗、阴影效应、多径衰落、抽头延时线模型、多普勒扩展;频率平坦衰落模型和频率选择性衰落模型。

3.1 基本概念

信道是发射天线与接收天线之间传输信号的介质,如图3-1所示。

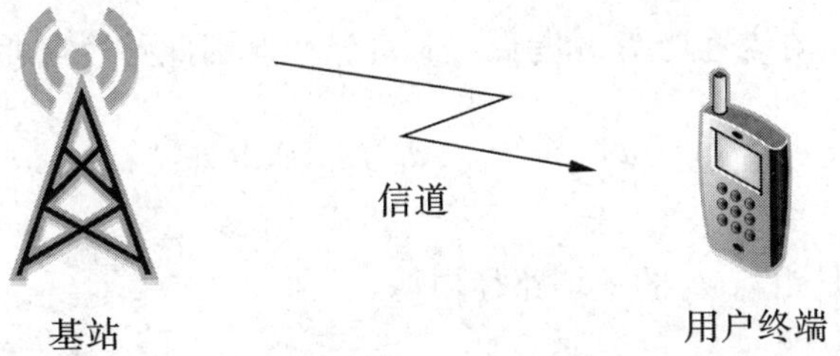

图3-1　信道模型

无线信号从发送端天线传输到接收端天线时,它的特性会发生变化。这些特性取决于两个天线之间的距离、信号传输的路径以及传输路径的周围环境(建筑物和其他物体)。如果事先知道发送端和接收端之间的传输媒介模型,结合发送信号,就可以获得接收信号的包络波形。这个传输媒介模型就是信道模型。

发送信号功率分布和信道的脉冲响应卷积可以得到接收信号功率分布。时域的卷积等于频域的乘积。因此,发送信号 x 经过信道 H 得到 y:

$$y(f)=H(f)x(f)+n(f) \tag{3.1}$$

其中 $H(f)$ 是信道状态信息,$n(f)$ 是噪声。注意,x,y,H 和 n 都是信号频率 f 的函数。

信道状态信息主要取决于路径损耗、阴影效应和多径衰落。

3.1.1 路径损耗

接收机和发送机之间没有视线障碍物的直视信道是最简单的信道。信号能量在传输过程中会衰减。对于直视(LOS)信道,接收功率由下式给出:

$$P_r = P_t \left[\frac{\sqrt{G_l}\lambda}{4\pi d} \right]^2 \tag{3.2}$$

其中 P_t 是发射功率,G_l 是发射天线增益和接收天线增益的乘积,λ 是波长,d 是距离。理论上讲,损耗功率与距离的平方成正比。而在实际当中,损耗功率与距离的三次方或四次方成正比。

地面会引起信号波反射再发送给接收机。经过地面反射的信号波到达发射机有可能会有 180°的相移,使得净接收功率减少。两条路径损耗近似可以表示为:

$$P_r = P_t \frac{G_t G_r h_t^2 h_r^2}{d^4} \tag{3.3}$$

这里 h_t 和 h_r 分别是发射机和接收机的天线高度。注意,这个公式和公式(3.2)有三个主要的区别:第一,天线高度对功率损耗会有影响;第二,波长被忽略;第三,距离指数为 4。一般来说,路径损耗公式可以简化为:

$$P_r = P_t P_0 \left(\frac{d_0}{d} \right)^{\alpha} \tag{3.4}$$

其中 P_0 是距离为 d_0 的信号功率,α 是路径损耗指数。因此路径损耗也可以表示为:

$$P_{\text{loss}}(d)_{\text{dB}} = \overline{P_{\text{loss}}}(d_0) + 10\alpha \log\left(\frac{d_0}{d} \right) \tag{3.5}$$

这里 $\overline{P_{\text{loss}}}(d_0)$ 是距离 d_0 的平均路径损耗。

3.1.2 阴影效应

如果在传输路径中有建筑物或树木等物体,它们会吸收、反射、散射和衍射传输信号,使得部分信息丢失。如果基站天线是光源,传输路径中的建筑会在接收天线上形成阴影,因此这种现象叫作阴影效应(图 3-2)。

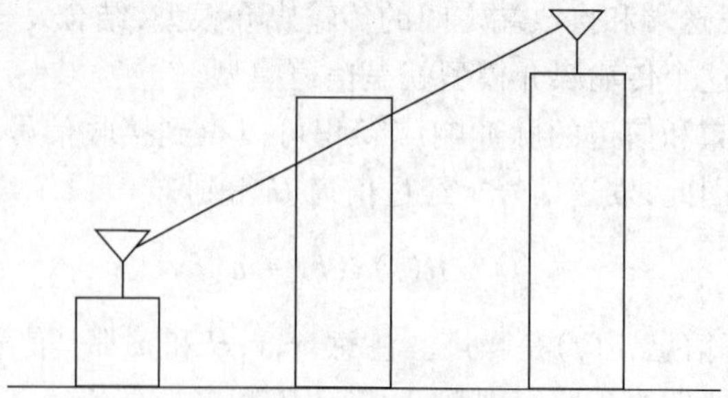

图 3-2 阴影效应

经过阴影效应的净路径损耗是:

$$\overline{P_{\text{loss}}(d)}_{\text{dB}} = \overline{P_{\text{loss}}}(d_0) + 10\alpha \log\left(\frac{d}{d_0} \right) + \chi \tag{3.6}$$

这里，χ 是具有标准偏差 σ 的正态（高斯）分布随机变量（以 dB 为单位），χ 代表阴影效应的影响。受阴影效应的影响，距离发射机相同距离 d 点处所接收的功率可以是不同的，但是符合对数正态分布。

3.1.3 多径衰落

信号传输路径周围的物体会反射信号。因此接收端也会接收到其中一部分反射波。每个反射信号波的路径不同，因此它们的幅度和相位不同（图 3-3）。

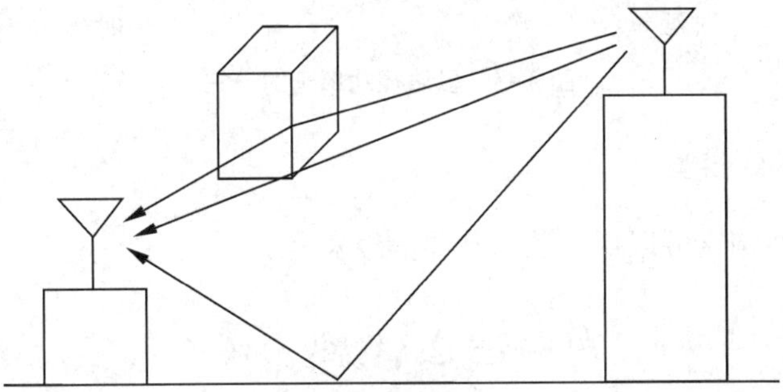

图 3-3 多径衰落

由于相位不同，多路径信号会使得接收功率增加或减小。略微改变位置也可能导致信号相位的显著差异。影响信道状态信息的的三个因素如图 3-4 所示。粗虚线表示路径损耗；细虚线表示符合对数正态分布的阴影衰落；实线粗线表示多径衰落。注意，由多径衰落引起的强度变化在信号波长范围内的较远处会发生变化。

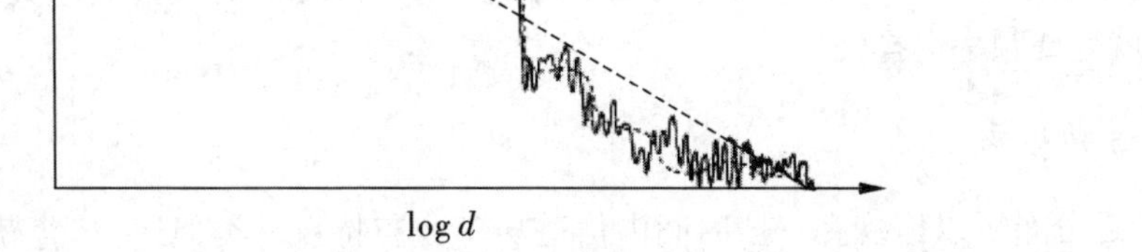

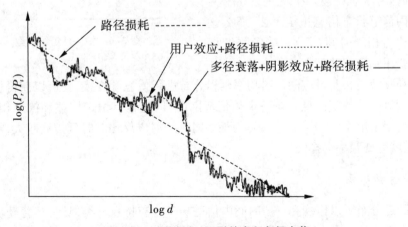

图 3-4 路径损耗、阴影效应和多径衰落

因为路径长度不同，接收端在不同时间里会收到发送机发送的单个脉冲信号的多个副本信号，如图 3-5 所示。

可以被忽略掉的接收信号最大时延叫做最大时延扩展 τ_{max}。大数值 τ_{max} 意味着这是高度弥散信道。通常我们不用最大时延扩展 τ_{max} 而使用均方根（rms）值。

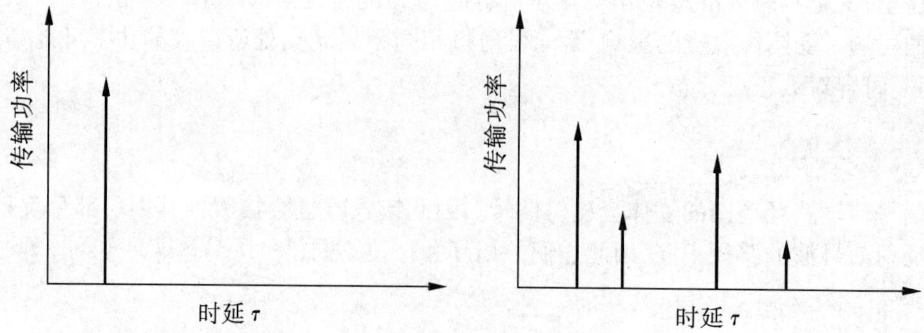

图 3-5 多路径功率延时分布

3.1.4 抽头延迟线模型

多径信道脉冲响应可以用离散的脉冲信号来表示：

$$h(t,\tau) = \sum_{i=1}^{M} c_i(t)\delta(\tau - \tau_i) \tag{3.7}$$

注意，脉冲响应 h 随时间 t 变化。系数 $c_i(t)$ 也随时间变化，上述模型中有 M 个系数。抽头延时线模型是通过 M 个抽头延时线来表示。例如，图 3-5 所示的信道可以用 4 抽头延时线模型表示，如图 3-6 所示。

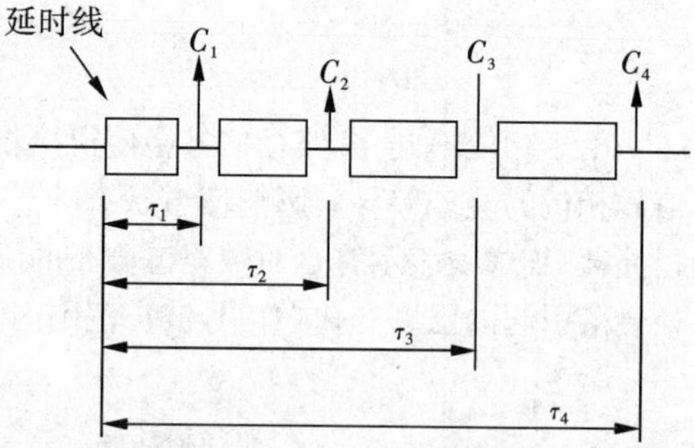

图 3-6 抽头延迟线模型

如果发送机、接收机或信道中的其他物体移动，则信道状态就会发生变化。如果在一段时间内信道状态信息恒定不变，那么这段时间被称为相干时间（T_c）。精确测量相干时间需要用到自相关函数。

时域和频域是相互对应的。对功率延迟分布进行傅里叶变换，就可以获得信道时域状态信息所对应的频率信息。在一定频带范围内，信道具有恒定增益和线性相位，这个频带被称为相干带宽（B_c）。相干带宽与最大时延扩展成反比。时延扩展越大，相干带宽越小，信道就呈现频率选择性。

3.1.5 多普勒扩展

功率延迟分布可以体现在一段时间内信道功率分布特点。类似地，多普勒功率谱是仅传输一个频率 f 信号的信道功率分布。功率延迟分布是由多径引起的，但是多普勒频谱是由传输路径上的物体运动引起的。多普勒频谱是 $(f-f_D, f+f_D)$，其中 f_D 是最大多普勒扩展（B_D）。

相干时间和多普勒扩展是成反比的：

$$相干时间 \approx \frac{1}{多普勒扩展} \tag{3.8}$$

因此,如果发送机、接收机或它们中间的物体移动得非常快,则多普勒扩展大,相干时间小,即信道变化快。

3.2 无线信道

"无线"表示用电磁波或射频(RF)来传输信息。无线通信的距离可以长也可以短。此外,它通常用于电线不可用的情况。下一代无线通信用户需要更高传输速率和更好传输质量,但频带资源是有限的。因此,了解无线衰落信道模型的特性是非常必要的。

无线通信包括点对点通信,点对多点通信,多点对多点通信,广播。图3-7展示了这些不同无线通信方式。无线设备比如手机,无线路由器,具有Wi-Fi功能的笔记本电脑以及调幅或调频无线电设备,通常使用9 kHz至300 GHz的频率。然而,在很多国家这个频率范围是公共资源,不同频带的应用不同。例如,普通GSM手机的频带是900 MHz或1800 MHz。飞行员与机场控制塔台进行通信的常规频段也约为900 MHz。如果乘客在起飞或着陆时使用手机,就可能干扰飞行员与控制塔台之间的通信。因此,有效利用有限的信道资源是非常有意义的。

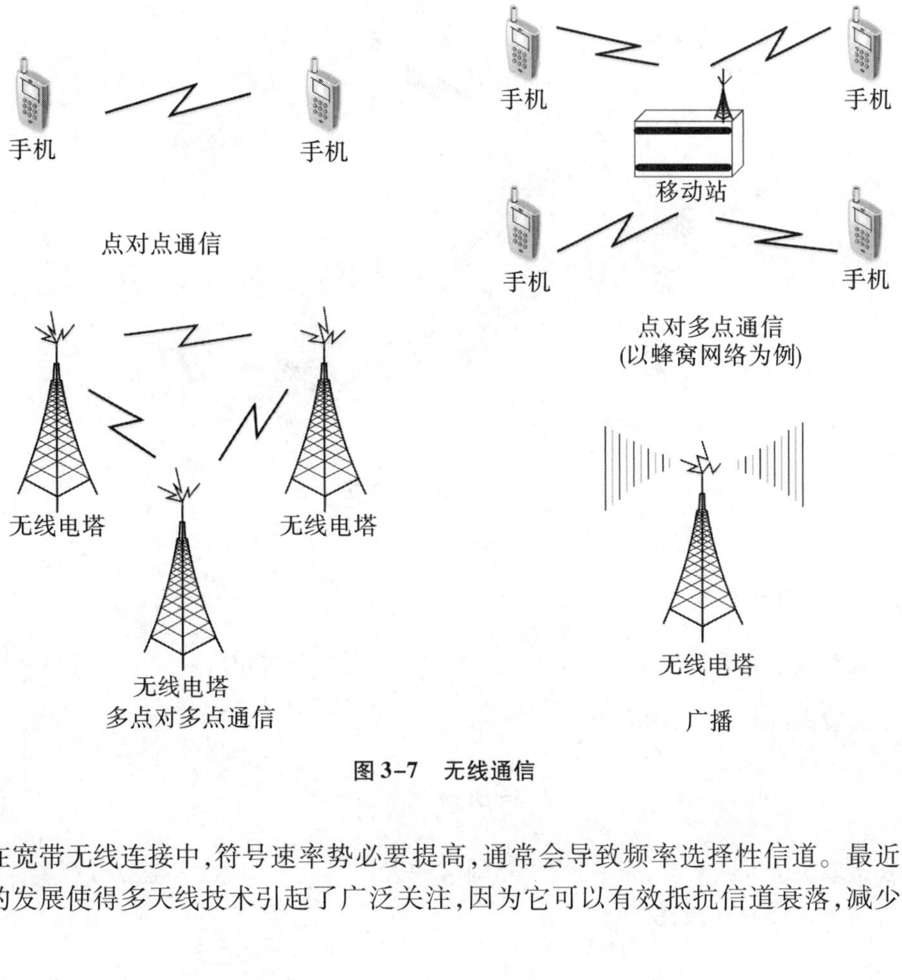

图3-7 无线通信

在宽带无线连接中,符号速率势必要提高,通常会导致频率选择性信道。最近通信技术的发展使得多天线技术引起了广泛关注,因为它可以有效抵抗信道衰落,减少信道干扰的影响。

3.2.1 衰落

衰落是指调制信号通过某个传播介质时能量发生衰减。在无线通信中,衰落是由多径传播或阴影效应引起的。

图3-8显示建筑物、云、树或飞机等障碍物可以反射、散射或衍射传输信号。非直视路径的传输波通常会被这些障碍破坏。因此,直视路径的接收信号通常比非直视路径的信号更强。

不同时间、频率和位置都会影响衰落,而常见的衰落信道模型包括两个:大规模衰落和小规模衰落。大规模衰落,也叫做衰减或路径损耗,体现信号在长距离或平均时间内的特性。相反,小规模衰落是发射信号的幅度或功率在短距离或短时间内的快速变化。衰减以及包括平坦衰落、频率选择性衰落、缓慢衰落、快速衰落、瑞利衰落、瑞森衰落等一些衰落信道都需要我们考虑和研究。

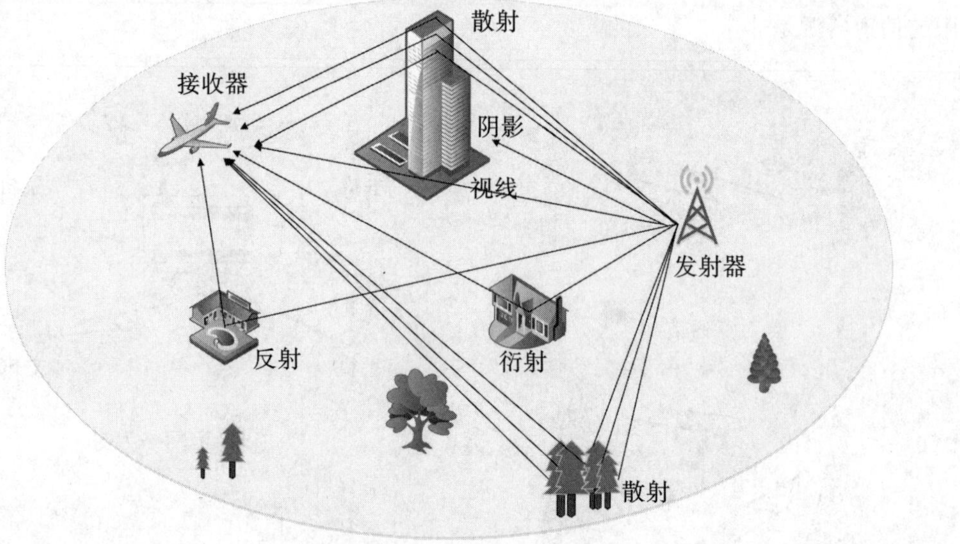

图3-8 衰落和多径

3.2.2 大规模衰落模型

大规模衰落或衰减表明在自由空间环境中由反射、散射和衍射所造成的传播路径损耗。距离发送机$\hat{d}$的平均功率为

$$P_{\hat{d}} = \beta(\hat{d}/\hat{d}_0)^{-v} P_t \tag{3.9}$$

其中β是取决于天线增益、频率、波长等因素的大规模衰落参数,$\hat{d}_0$是参考距离,P_t是平均发射功率,v是路径损耗指数。

在一个自由空间中,路径损耗指数通常等于2。如果传输路径有障碍物,如图3-8所示,路径损耗指数大于2。距离发送端或基站相同距离但不同位置的手机测量的$P_{\hat{d}}$可能不一样,因为障碍物会随机通过阴影效应影响路径损耗。大规模衰落中平均功率的概率密度函数(pdf)是高斯分布。图3-9对应公式(3.9)的大规模衰落模型,其中载波频率是1800 MHz,参考距离$\hat{d}_0$是100 m。在该模型中考虑了三种类型的情况:v在自由空间蜂窝无线通信系统为2,在城市蜂窝无线通信中为3,在阴影效应下的城市蜂窝无线通信是5。水平轴是对数距离,垂直轴是以dB为单位的路径损耗。在没有阴影影响时,路径损耗随v增加。但是随机阴影效应会对路径损耗产生影响。

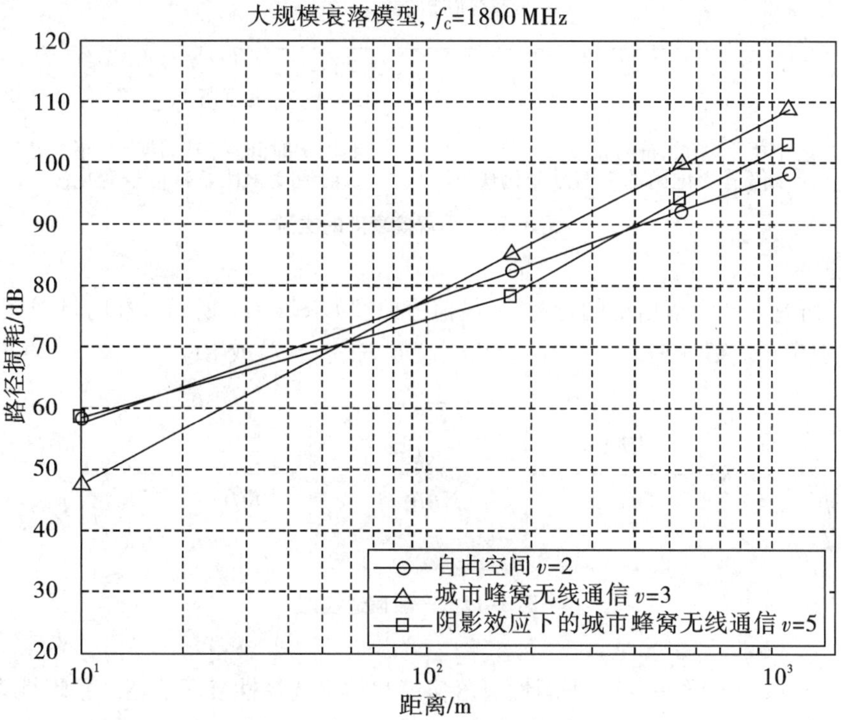

图3-9　大规模衰落模型

3.2.3　小尺度衰落模型

小尺度衰减或衰落表明接收机在小范围内移动时发射信号功率快速变化。它的衰落是由传输信号多径波引起的,这些波在不同时间到达接收机。如前所述,散射、衍射和反射都可以产生多径波,也就是多径衰落。

基于信号参数(如带宽、符号周期等)和信道参数(如均方根时延扩展和多普勒扩展)之间的关系,传输信号会经历不同的衰落。多径时延扩展会导致时间扩散和频率选择性衰落,多普勒扩展会导致频率扩散和时间选择性衰落。

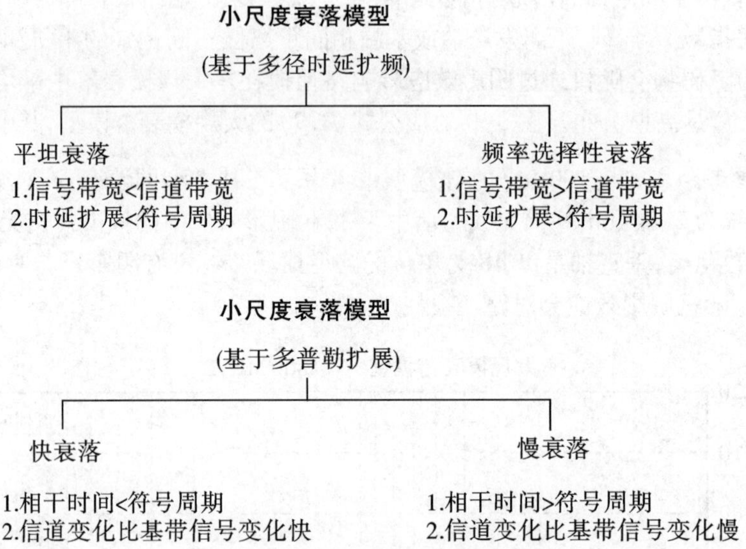

图 3-10　小规模衰落的类型

为了研究不同衰落信道的行为，一般信道模型如图 3-11 所示。在时域或频域中，信源，信道响应和信宿分别为 $s(t)$ 或 $s(f)$，$h(t)$ 或 $H(f)$，$r(t)$ 或 $R(f)$。

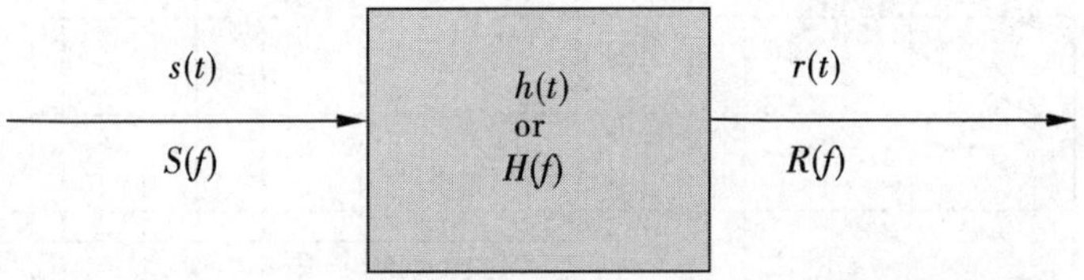

图 3-11　一般信道模型

图 3-12(a)(b)分别表示平坦性衰落信道和频率选择性衰落信道。平坦性衰落信道是相干带宽大于信源带宽。这里的相干表示信道幅度保持不变所需要的最小频率或时间。另外，如图 3-12(b)所示，信源带宽大于相干带宽，则接收信号会失真。$s(f)$ 是频域中的符号带宽，这样会存在符号间干扰(ISI)。因此，信道带宽 $H(f)$ 应等于或大于信号相干带宽从而避免 ISI。

如果信道的相干时间较长，则会发生慢衰落。前面所说的阴影效应会导致信道慢衰落。信道变化的速率比传输信号的速率慢得多。信道幅度和相位在一个或几个频率带宽内被认为是静态的。另外，如果相干时间小于符号周期，则会发生快速衰落。这时信道振幅和相位不再是静态的，并且在符号周期内快速变化。

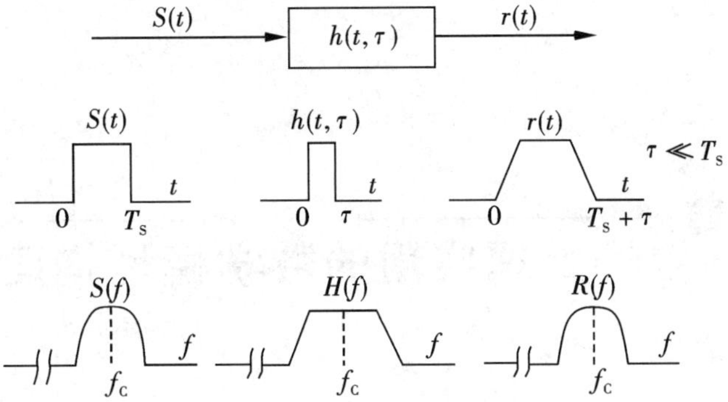

(a)平坦性衰落信道特性

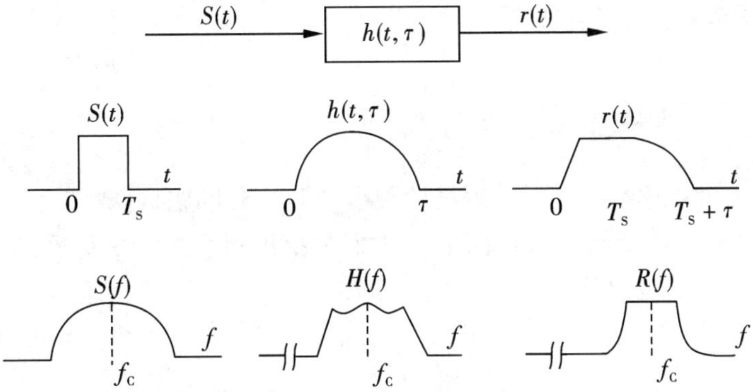

(b)频率选择性信道特性

图3-12 平坦性衰落信道特性和频率选择性衰落信道特性

总之,线性时变脉冲响应可以构建平坦性或频率选择性衰落信道,慢衰落或快衰落信道。如果满足 $B_s<B_c$ 和 $T_s<\sigma_\tau$,则信号经历平坦衰落;如果满足 $B_s<B_c$ 和 $T_s<\sigma_\tau$ 或者满足 $T_s<10\sigma_\tau$,则信号经历频率选择性衰落。其中 T_s 是带宽的倒数(符号周期),σ_τ 是均方根延时扩展。但是,我们很难用理想脉冲响应模型模拟实际环境。因此,瑞利衰落信道模型,瑞森衰落信道模型等是研究接收信号性能的重要模型。

平坦性衰落信道主要分两种情况。如果发射机和接收机之间是非直视路径,则它是瑞利衰落信道。否则,它是一个瑞森衰落信道。公式(3.10)和公式(3.11)分别给出了瑞利随机变量或瑞森随机变量的概率密度函数(PDFs):

$$f_{\text{Rayleigh}}(r)=\frac{r}{\sigma^2}\exp\left(\frac{-r^2}{2\sigma^2}\right),r\geqslant 0 \tag{3.10}$$

$$f_{\text{rician}}(r)=\frac{r}{\sigma^2}\exp\left(\frac{-(r^2+D^2)}{2\sigma^2}\right)I_0\left(\frac{D_r}{\sigma^2}\right),r\geqslant 0,D\geqslant 0 \tag{3.11}$$

其中 r 是接收信号幅度,σ^2 是平均功率,D 表示直视路径信号的峰值振幅,$I(\cdot)$是第一类修正贝赛尔函数,$I_0(\cdot)$是 $I(\cdot)$的零阶,如果是非直视信道,瑞森分布近似瑞利分布,这时 D 接近零。

第 4 章 信号处理应用的数学模型构建和实现

正则方程组可以用来描述很多信号处理应用,如何求解正则方程组以及如何进行硬件实现是主要面临的问题之一。本章将详细介绍和对比求解正则方程组的方法:直接法和迭代法,以及典型算法硬件实现。

4.1 求解正则方程组

许多信号处理应用都是实时求解线性最小二乘(LS)问题;这些应用包括自适应天线阵列,多用户检测,MIMO 检测,回波消除,均衡,系统识别等。线性 LS 问题等同于求一组线性方程或正则方程组的解。

$$\boldsymbol{Ax} = \boldsymbol{b} \tag{4.1}$$

其中 $\boldsymbol{A}$ 是一个 $N\times N$ 的对称正定矩阵。$\boldsymbol{x}$ 和 $\boldsymbol{b}$ 是 $N\times 1$ 的向量。矩阵 $\boldsymbol{A}$ 和向量 $\boldsymbol{b}$ 已知,求解估计向量 $\boldsymbol{x}$。通常求解准确值的方法是

$$\boldsymbol{x} = \boldsymbol{A}^{-1}\boldsymbol{b} \tag{4.2}$$

其中 $\boldsymbol{A}^{-1}$ 是矩阵 $\boldsymbol{A}$ 的逆矩阵。众所周知,矩阵求逆的计算量与矩阵 N 的大小有关,求逆的复杂度是 $O(N^3)$。标准数学计算软件,例如 Matlab,通过 LAPACK 库的函数可以解决这个问题。从数值的角度来看,矩阵求逆的最好方法不是直接计算而是通过求解合适的方程组获得。而实时求解 LS 问题的最好方法就是求解方程组。目前众多有效方法主要可以分为两大类:直接法和迭代法。直接法是通过有限数量的预定操作来计算方程组的精确解。相对应的,迭代法是连续生成一系列逼近最优解的近似解。

这些方法也可以用于矩阵求逆 $\boldsymbol{A}^{-1}$。设 $\boldsymbol{AX}=\boldsymbol{I}$,其中 $\boldsymbol{I}$ 是一个 $N\times N$ 单位矩阵,需要计算 $N\times N$ 的矩阵 $\boldsymbol{X}=\boldsymbol{A}^{-1}$,我们得到 N 个方程组

$$\boldsymbol{AX}_{:,n} = \boldsymbol{I}_{:,n} \quad n = 1,\cdots,N. \tag{4.3}$$

通过求解这 N 个系统方程组,可以得到 $\boldsymbol{A}^{-1}$。接下来,我们将介绍一些解决线性系统问题的直接法和迭代法。

4.1.1 直接法

直接法,例如高斯消元、$\boldsymbol{LU}$ 分解、Cholesky 分解、QRD 等是在有限序列的预定操作之后,可以得到方程组(4.1)的精确解。直接法的核心思想是将一般的方程组简化为上三

角形或下三角形形式,其解与原方程的解相同。方程组可以通过向前或者向后替换来求解,并且能够提供高精度的解。

高斯消元法的基本思想是采取适当的线性组合修改原方程(4.1)从而得到等效的三角形系统。具体地,就是采用行变换把方程组(4.1)变成上三角方程 $\boldsymbol{Ux}=\boldsymbol{y}$,其中 $\boldsymbol{U}$ 是一个 $N\times N$ 维的上三角矩阵,$\boldsymbol{y}$ 是一个 $N\times 1$ 维的向量。求解上三角形方程 $\boldsymbol{Ux}=\boldsymbol{y}$ 可以用向后替换操作。高斯消元法的复杂度是 $2N^3/3$,复杂度包括乘法、除法和加法。除了高复杂度之外,高斯消元法的主要缺点是(4.1)式中的右侧向量 $\boldsymbol{b}$ 需要参与消元过程,只有在右侧向量已知的情况下才能进行消元。

$\boldsymbol{LU}$ 分解可以被看作是高斯消元法的高级代数描述。它将系统方程组(4.1)中的矩阵 $\boldsymbol{A}$ 分解成一个乘积,$\boldsymbol{A}=\boldsymbol{LU}$,其中 $\boldsymbol{L}$ 是单位下三角矩阵,其所有主对角元素都等于1,$\boldsymbol{U}$ 是上三角矩阵。因此,可以连续地通过向前替换计算下三角矩阵 $\boldsymbol{Ly}=\boldsymbol{b}$ 和向后替换计算上三角矩阵 $\boldsymbol{Ux}=\boldsymbol{y}$,求得向量 $\boldsymbol{x}$。与高斯消元法相比,$\boldsymbol{LU}$ 分解的优点是可以不依靠右侧向量 $\boldsymbol{b}$ 来修改(或分解)矩阵。如果已知方程组(4.1)的解时,不分解矩阵就可以求解具有相同左侧矩阵 $\boldsymbol{A}$ 的其他矩阵系统。因此,用相同左侧矩阵求解系统方程的复杂性会显著降低。在实际应用中,$\boldsymbol{LU}$ 分解的应用很广泛。然而,$\boldsymbol{LU}$ 分解技术非常复杂,是 $2N^3/3$,包括乘法、除法和加法。

因为正则方程(4.1)中的系数矩阵 $\boldsymbol{A}$ 是对称正定的,Cholesky 可以对它进行分解。Cholesky 分解与高斯消元法密切相关。Cholesky 分解把正定系数矩阵 $\boldsymbol{A}$ 分解成 $\boldsymbol{A}=\boldsymbol{U}^{\mathrm{T}}\boldsymbol{U}$,其中 $\boldsymbol{U}$ 是主对角线元素均为正数的上三角矩阵。因此,方程组(4.1)可以被写为 $\boldsymbol{U}^{\mathrm{T}}\boldsymbol{Ux}=\boldsymbol{b}$。设 $\boldsymbol{y}=\boldsymbol{Ux}$,可以得到一个下三角形系统 $\boldsymbol{U}^{\mathrm{T}}\boldsymbol{y}=\boldsymbol{b}$。分别通过向前和向后替换的方式依次求解下三角形系统方程 $\boldsymbol{U}^{\mathrm{T}}\boldsymbol{y}=\boldsymbol{b}$ 和上三角形系统方程 $\boldsymbol{Ux}=\boldsymbol{y}$,可以求得解向量 $\boldsymbol{x}$。与高斯消元法相比,Cholesky 分解法的优点是只需要一半操作数和占用一半内存空间。此外,由于正定矩阵 $\boldsymbol{A}$ 是非奇异矩阵,Cholesky 分解可以保证数值稳定性。然而,Cholesky 分解的复杂度是 $N^3/3$ 次运算,包括乘法和除法运算。对于实时硬件实现来说,运算量仍然太高,尤其当系统规模 N 很大的时候。

QRD 以数值稳定性而闻名,并广泛用于许多应用。它用和 $\boldsymbol{LU}$ 分解相同的方式求解方程组(4.1);它把系数矩阵 $\boldsymbol{A}$ 转换为 $\boldsymbol{A}=\boldsymbol{QR}$,其中 $\boldsymbol{Q}$ 是正交矩阵,$\boldsymbol{R}$ 是上三角矩阵。正交矩阵有 $\boldsymbol{QQ}^{\mathrm{T}}=\boldsymbol{Q}^{\mathrm{T}}\boldsymbol{Q}=\boldsymbol{I}$ 和 $\boldsymbol{Q}^{-1}=\boldsymbol{Q}^{\mathrm{T}}$ 的性质,其中 $\boldsymbol{I}$ 是 $N\times N$ 单位矩阵。因此,系统方程(4.1)可以变换成上三角系统 $\boldsymbol{Rx}=\boldsymbol{Q}^{\mathrm{T}}\boldsymbol{b}$,然后用向后替换操作获得解向量 $\boldsymbol{x}$。

$\boldsymbol{QR}$ 等同于计算一组向量的正交基。计算 QRD 时有多很多方法,例如 Householder 反射和 Givens 旋转。反射和旋转易于构造,通过旋转一定角度或反射平面可以把向量中的某些元素置换为0。Householder 反射通过置零法消去向量中的所有元素(除了第一个元素之外)是非常有效的。Givens 旋转在向量中置零有很多选择,包括第一个元素。因此,Givens 旋转是 QRD 变形转换的主要选择。

如果使用 Givens 旋转,则 QRD 非常适合硬件实现,因为它可以对排列成三角形结构的脉冲阵列架构进行并行处理和流水线处理。这种阵列中的每个处理单元都有自己的本地存储器,并且只连接到最近的单元。阵列的特殊结构使得阵列里的数据以流水线形式工作。这种简单且高度并行的阵列使得数据流水线传输,从而提升了传输吞吐量。因此,它非常适合实现复杂的信号处理算法,特别是对于实时和高数据带宽的实现。但是,这种三角阵列结构算法的复杂性与系统大小密切相关,处理单元数目 $(N^2+N)/2$ 随着矩阵尺寸 N 的增大而急剧增加,这使得应用脉冲阵列的直接硬件设计是非常昂贵的,例如自适应波束形成,需要多个芯片解决,而非一个芯片。因此,三角形架构只适用于尺寸较小的矩阵。

另外,我们可以用一些较简单的架构阵列来解决大尺寸矩阵的问题。在文献[17]中,直接映射三角型架构脉冲阵列可以得到 QRD 线性结构脉冲阵列;线性阵列的每个处理单元对应三角阵列中每行的处理单元。这种线性阵列把处理单元数量减少到 N。然而,文献[17]中的线性阵列处理单元比三角形阵列的处理单元复杂得多,因为它是通过合并三角形阵列的对角线单元和非对角线单元的函数而获得的。而且,并不是所有的处理单元都能 100% 被利用。另一种线性脉冲阵列是通过在三角形脉冲阵列上采用折叠和映射的方法获得的。这种线性阵列保留了三角形脉冲阵列的本地互连,并且只需要 M 个处理单元($N=2M+1$ 是矩阵大小)。这些处理单元具有与三角形阵列相似的复杂性。而且,这些处理单元被 100% 利用。文献[15]提出了合并相似三角形阵列处理单元,增加存储器块和使用控制逻辑来调度块之间的数据移动,从而避免了三角形结构阵列。与线性架构相比,组合型架构通过较长的延迟,降低了复杂度。因此,我们需要在性能和复杂性之间对选取线性架构阵列和组合型架构进行权衡。组合型架构需要更多的控制逻辑,更长的延迟,但与三角形结构脉冲阵列相比,所需处理单元的数量显著减少。

典型 Givens 旋转包含平方根、除法和乘法运算,这对于硬件实现来说是极其昂贵的。研究在硬件上使用 Givens 旋转的 QRD 高效脉冲阵列实现已经有了很多成果,这些成果可以分为三种主要类型。使用 Givens 旋转的 QRD 的第一种类型是坐标旋转数字计算方法(CORDIC)。CORDIC 是计算三角函数的迭代技术,如正弦和余弦。CORDIC 很简单,它只需要移位和加法操作,不需要任何乘法、除法和平方根操作。因此,它非常适合于定点型硬件的实现。然而,由于 CORDIC 中用定点型表示的动态范围有限,因此在精度相同的情况下,字长要求比浮点型要大得多。此外,CORDIC 算法有许多子旋转,导致较大误差。

使用 Givens 旋转的 QRD 第二种类型是基于免平方根的 Givens 旋转或平方 Givens 旋转(SGR),这种类型不需要进行平方根操作并减少了一半数量的乘法操作。与典型 Givens 旋转以及 CORDIC 技术相比,SGR 方法有这样几个优点。第一,它很简单,不需要平方根操作。但是,与需有平方根运算的经典 Givens 旋转相比,它数值准确性变差。第二,它比 CORDIC 运算时间要快得多。与 CORDIC 的硬件设计相比,SGR 的硬件面积缩小了大约两倍,并且延迟仅为 CORDIC 硬件实施的 66%。此外,在相同精度的情况下,浮点型 SGR 所需要的比特数是定点型 CORDIC 的 80%。

使用 Givens 旋转的 QRD 第三种类型是基于对数数字系统(LNS)的。在 LNS 算法中,传统数字系统的平方根运算变成简单的位移运算,而传统数字系统的乘法和除法运算分别成为加法和减法运算。然而,简单的加法或减法操作运算在 LNS 算法却需要较大开销。因此,即使基于 LNS 算法的 Givens 旋转操作不需要乘法、除法和平方根,它的加法操作在硬件实现上也非常复杂。另外,基于 LNS 的 QRD 需要许多转换操作才能适合传统数值系统设计。

一般来讲,大规模系统使用脉冲阵列 QRD。对于小尺寸矩阵,有一些比 QRD 更快,硬件实施效率更高,同时能提供足够数值稳定性的替代算法。矩阵求逆的一种直接的方法是解析方法。例如,使用解析方法对 2×2 矩阵求逆计算如下:

$$\boldsymbol{B}^{-1}=\begin{bmatrix} a & b \\ c & d \end{bmatrix}^{-1}=\frac{1}{ad-bc}=\begin{bmatrix} d & -b \\ -c & a \end{bmatrix} \tag{4.4}$$

对于小尺寸矩阵,解析方法的复杂性明显小于 QRD。因此,解析方法对于小尺寸矩阵求逆是相当有效的,例如等式(4.4)中的 2×2 矩阵。然而,解析方法的复杂性随着矩阵尺寸 N 的增加而迅速增长,这使得它仅适用于小尺寸的矩阵。此外,直接解析矩阵求逆对字节有限长度误差较敏感。即使对 4×4 矩阵,直接解析方法也是不稳定的,因为计算中包含的大量减法可能会引入数值抵消。也就是说,由于数值表示没有达到足够的精度(或字节数),由于字节长度引起的结果错误会显著影响接收端的性能。

在文献[22]中,提出了一种称为分块矩阵求逆分析(BAMI)来计算复数矩阵的逆运算。它将矩阵分成四个较小的矩阵,然后局部计算逆矩阵。例如,要计算一个 4×4 的矩阵 $\boldsymbol{B}$,它首先被分成 4 个 2×2 的子矩阵。

$$\boldsymbol{B}=\begin{bmatrix} \boldsymbol{B}_1 & \boldsymbol{B}_2 \\ \boldsymbol{B}_3 & \boldsymbol{B}_4 \end{bmatrix} \tag{4.5}$$

因此,矩阵 $\boldsymbol{B}$ 求逆可以用分析方法(4.4)求这些 2×2 矩阵的逆来获得,即:

$$\boldsymbol{B}^{-1}=\begin{pmatrix} \boldsymbol{B}_1^{-1}+\boldsymbol{B}_1^{-1}\boldsymbol{B}_2(\boldsymbol{B}_4-\boldsymbol{B}_3\boldsymbol{B}_1^{-1}\boldsymbol{B}_2)^{-1}\boldsymbol{B}_3\boldsymbol{B}_1^{-1} & -\boldsymbol{B}_1^{-1}\boldsymbol{B}_2(\boldsymbol{B}_4-\boldsymbol{B}_3\boldsymbol{B}_1^{-1}\boldsymbol{B}_2)^{-1} \\ -(\boldsymbol{B}_4-\boldsymbol{B}_3\boldsymbol{B}_1^{-1}\boldsymbol{B}_2)^{-1}\boldsymbol{B}_3\boldsymbol{B}_1^{-1} & (\boldsymbol{B}_4-\boldsymbol{B}_3\boldsymbol{B}_1^{-1}\boldsymbol{B}_2)^{-1} \end{pmatrix} \tag{4.6}$$

因为减法的次数较少了, BAMI 方法比直接解析方法更稳定,保持同样的精度所需要的比特数较少。因此,在求解小尺寸矩阵时,例如 4×4 矩阵,BAMI 为典型 QRD 提供了一个很好的可选方案。然而,对于 2×2 和 3×3 矩阵,直接解析法是首选。

Sherman–Morrison 方程是矩阵求逆引理的一个特例,它简单计算一系列连续矩阵的逆矩阵,其中这一系列矩阵中连续两个矩阵间有一个微小的摄动区别。这个摄动必须是秩 1 更新形式,例如 $\boldsymbol{u}\boldsymbol{v}^{\mathrm{H}}$,其中 $\boldsymbol{u}$ 和 $\boldsymbol{v}$ 是向量。给定 $\boldsymbol{A}^{-1}$,Sherman–Morrison 公式表示为:

$$(\boldsymbol{A}^{-1}+\boldsymbol{u}\boldsymbol{v}^{\mathrm{H}})^{-1}=\boldsymbol{A}^{-1}-\frac{(\boldsymbol{A}^{-1}\boldsymbol{u}\,\boldsymbol{v}^{\mathrm{H}})\boldsymbol{A}^{-1}}{1+\boldsymbol{v}^{-1}\boldsymbol{A}^{-1}\boldsymbol{u}} \tag{4.7}$$

因此,对于一系列矩阵 $\boldsymbol{A}(i)=\boldsymbol{A}(i-1)+\boldsymbol{u}(i)\boldsymbol{u}^{\mathrm{H}}(i)$,例如,自相关矩阵,其中 i 是时间索引, $\boldsymbol{u}(i)$ 是输入向量,可以通过 Sherman–Morrison 公式很容易地计算出 $\boldsymbol{A}^{-1}(i)$,例如

$$\begin{aligned} \boldsymbol{A}^{-1}(i) &= [\boldsymbol{A}(i-1)+\boldsymbol{u}(i)\boldsymbol{u}^{\mathrm{H}}(i)]^{-1} \\ &= \boldsymbol{A}^{-1}(i-1)-\frac{\boldsymbol{A}^{-1}(i-1)\boldsymbol{u}^{\mathrm{H}}(i)\boldsymbol{A}^{-1}(i-1)}{1+\boldsymbol{u}^{\mathrm{H}}(i)\boldsymbol{A}^{-1}(i-1)\boldsymbol{u}(i)} \end{aligned} \tag{4.8}$$

这种矩阵求逆的方法被广泛使用,例如在 MIMO 系统中。然而,等式(4.8)需要大量的乘法和除法,硬件实现难以操作。在文献[24]中,通过适当的缩放操作,可以把等式(4.8)中的除法转换为乘法,例如

$$
\begin{aligned}
\widetilde{\boldsymbol{A}}^{-1}(i) &= [\alpha(i-1)+\boldsymbol{u}^{\mathrm{H}}(i)\widetilde{\boldsymbol{A}}^{-1}(i-1)\boldsymbol{u}(i)]\left[\widetilde{\boldsymbol{A}}(i-1)+\frac{\boldsymbol{u}(i)\boldsymbol{u}^{\mathrm{H}}(i)}{\alpha(i-1)}\right]^{-1} \\
&= \widetilde{\boldsymbol{A}}^{-1}(i-1)[\alpha(i-1)+\boldsymbol{u}^{\mathrm{H}}(i)\widetilde{\boldsymbol{A}}^{-1}(i-1)\boldsymbol{u}(i)]- \\
&\quad [\widetilde{\boldsymbol{A}}^{-1}(i-1)\boldsymbol{u}(i)\boldsymbol{u}^{\mathrm{H}}(i)\widetilde{\boldsymbol{A}}^{-1}(i-1)]
\end{aligned}
\tag{4.9}
$$

其中 $\widetilde{\boldsymbol{A}}^{-1}(i)=\alpha(i)\widetilde{\boldsymbol{A}}^{-1}(i)$,比例因子

$$
\alpha(i)=\alpha(i-1)[\alpha(i-1)+\boldsymbol{u}^{\mathrm{H}}(i)\widetilde{\boldsymbol{A}}^{-1}(i-1)\boldsymbol{u}^{\mathrm{H}}(i)] \tag{4.10}
$$

其中 $\alpha(0)=1$。然而,修改后的 Sherman-Morrison 方程(4.9)仍然非常复杂,其复杂度是 $O(N^3)$ 个乘法运算,乘法在硬件设计上是非常昂贵的。

直接法,例如高斯,Cholesky 和 QRD 等的复杂度是 $O(N^3)$ 个操作数,需要除法和乘法运算。改进的 Sherman-Morrison 方法大约需要 $O(N^2)$ 次乘法运算。因此,直接法很难用于实时信号处理和硬件实现。直接法在有限预设操作之后可以计算出确切的解。它们只有在执行完所有预定操作之后才能得出结果。因此,如果预设操作被提早结束,直接法就无法给出结果。此外,直接法求解大规模或稀疏线性方程组是异常复杂的。

4.1.2 迭代法

比较于直接法,迭代法分析大规模系统和非常稀疏系统非常有效。迭代法连续生成逐步精确的估计值 $\boldsymbol{x}(k)$ (k 是迭代次数),最终收敛到最优解,迭代法是通过矩阵向量乘法计算系数矩阵 $\boldsymbol{A}$。

求解正则方程组(4.1)可以转换为最小化二次函数:

$$
f(x)=\frac{1}{2}\boldsymbol{x}^{\mathrm{T}}\boldsymbol{A}\boldsymbol{x}-\boldsymbol{x}^{\mathrm{T}}\boldsymbol{b} \tag{4.11}
$$

通过设定 $\boldsymbol{x}=\boldsymbol{A}^{-1}\boldsymbol{b}$ 作为正态方程(4.1)解,$f(\boldsymbol{x})$ 的最小值是 $-\frac{1}{2}\boldsymbol{b}^{\mathrm{T}}\boldsymbol{A}^{-1}\boldsymbol{b}$。大多数迭代方法通过最小化函数 $f(\boldsymbol{x})$ 来迭代求解正态方程(4.1)。设定初始值是 $\boldsymbol{x}^{(0)}$,并生成一系列迭代结果 $\boldsymbol{x}^{(1)},\boldsymbol{x}^{(2)},\cdots$。在每个步骤(或迭代)中,选择满足条件 $f(\boldsymbol{x}^{(k+1)})\leqslant f(\boldsymbol{x}^{(k)})$,甚至满足 $f(\boldsymbol{x}^{(k+1)})<f(\boldsymbol{x}^{(k)})$ 的数值 $\boldsymbol{x}^{(k+1)}$ 。逐步接近 $f(\boldsymbol{x})$ 的最小值。经过多次迭代之后,如果 $\boldsymbol{x}^{(k)}$ 可以满足或者近似满足 $\boldsymbol{A}x^{(k)}=\boldsymbol{b}$ 则停止算法并把 $\boldsymbol{x}^{(k)}$ 作为系统(4.1)的解。

从 $\boldsymbol{x}^{(k)}$ 到 $\boldsymbol{x}^{(k+1)}$ 的计算步骤有两个要素:①选择一个方向向量 $\boldsymbol{p}^{(k)}$,表示 $\boldsymbol{x}^{(k)}$ 到 $\boldsymbol{x}^{(k+1)}$ 的行进方向;②在 $\boldsymbol{x}^{(k)}+\alpha^{(k)}\boldsymbol{p}^{(k)}$ 上选择一个点作为 $\boldsymbol{x}^{(k+1)}$,其中 $\alpha^{(k)}$ 是使的 $f(\boldsymbol{x}^{(k)}+\alpha^{(k)}\boldsymbol{p}^{(k)})$ 最小化的步长。选择 $\alpha^{(k)}$ 的过程称为线搜索。要选择一个合适 $\alpha^{(k)}$ 使得 $\boldsymbol{x}^{(k+1)}\leqslant\boldsymbol{x}^{(k)}$ 。若要确保这一条件则选择的 $\alpha^{(k)}$ 需要满足

$$f(\boldsymbol{x}^{(k+1)}) = \min f(\boldsymbol{x}^{(k)} + \alpha^{(k)} \boldsymbol{p}^{(k)}) \tag{4.12}$$

这个过程叫作精确线搜索,否则就是近似线搜索。

迭代法有两种主要类型:非定常迭代法和定常迭代法。非定常迭代法是近些年发展起来的,包括最速下降法,共轭梯度(CG)法等等。他们运算复杂,但是非常有效的。定常迭代法出现较早,操作简单,但不如非定常迭代法有效。线性系统有三种常见的定常迭代法: Jacobi,Gauss-Seidel 和 SOR 方法。

一个著名的迭代法是最速下降法。最速下降法是在负梯度方向进行线搜索:

$$\boldsymbol{p}^{(k)} = -\nabla f(\boldsymbol{x}^{(k)}) = \boldsymbol{b} - \boldsymbol{A}\,\boldsymbol{x}^{(k)} = \boldsymbol{r}^{(k)} \tag{4.13}$$

其中 $\boldsymbol{r}^{(k)}$ 是结果 $\boldsymbol{x}^{(k)}$ 的残差矢量,步长是

$$\alpha^{(k)} = \frac{(\boldsymbol{r}^{(k)})^{\mathrm{T}} \boldsymbol{r}^{(k)}}{(\boldsymbol{r}^{(k)})^{\mathrm{T}} \boldsymbol{A}\, \boldsymbol{r}^{(k)}} \tag{4.14}$$

最速下降法很容易编程,但它通常收敛缓慢。它收敛速度慢的主要是因为在平行或几乎平行的搜索方向上最小化 $f(\boldsymbol{x}^{(k)})$ 需要花费时间。最速下降法的复杂度很高,每次迭代都需要 $O(N^2)$ 个乘法、除法和加法运算。

共轭梯度算法是快速收敛最速下降法的简单版本。它是通过在包括先前所有搜索方向,例如

$$\boldsymbol{x}^{(k)} = \alpha^{(0)} \boldsymbol{p}^{(0)} + \alpha^{(1)} \boldsymbol{p}^{(1)} + \cdots \alpha^{(k-1)} \boldsymbol{p}^{(k-1)} \tag{4.15}$$

的超平面上最小化函数 $f(\boldsymbol{x}^{(k)})$,加速接近最优解的收敛速度,而不像最速下降法仅在下降梯度方向最小化 $f(\boldsymbol{x}^{(k)})$ 。由于共轭梯度算法的快速收敛性,长期被用于自适应滤波。然而,共轭梯度迭代算法的复杂度是 $O(N^2)$,其中包括除法、乘法和加法,这些操作对于实时信号处理来说太过复杂。

Jacobi 方法是最简单的迭代方法。它通过

$$x_n^{(k+1)} = (b_n - \sum_{p \neq n} A_{n,p} x_p^{(k)}) / A_{n,n} \tag{4.16}$$

从 $\boldsymbol{x}^{(0)}$ 更新下一个迭代 $\boldsymbol{x}^{(k+1)}$ 。其中 $A_{n,p}$ 、b_n 和 x_n 分别是系数矩阵 $\boldsymbol{A}$ 的第 (n,p) 个元素,向量 $\boldsymbol{b}$ 和 $\boldsymbol{x}$ 的第 n 个元素。Jacobi 方法具有以下优点: $\boldsymbol{x}^{(k+1)}$ 的所有元素都是彼此独立的,因此 $\boldsymbol{x}^{(k+1)}$ 的所有元素可以同时执行,Jacobi 方法本质上是平行操作。另一方面,如(4.16)所示,Jacobi 方法不用最新可用信息来计算 $x_n^{(k+1)}$ 。$\boldsymbol{x}^{(k)}$ 在获得下一个迭代数值 $\boldsymbol{x}^{(k+1)}$ 之前只能被覆写,因此,Jacobi 需要存储 $\boldsymbol{x}$ 的两个副本。如果系统规模 N 很大,$\boldsymbol{x}$ 的每个副本将占用大量的内存空间。此外,如果条件不能满足,Jacobi 通常可以通过置换行和列来实现矩阵 $\boldsymbol{A}$ 对角线元素是非零值从而避免(4.16)中的分母为 0。此外,Jacobi 方法并不总是收敛到最优解。但是,即使在收敛速度缓慢的情况下,Jacobi 方法在条件满足的情况下(例如,如果 $\boldsymbol{A}$ 是严格对角优势矩阵)也是可以保证收敛性的。

修正 Jacobi 迭代使它可以用最新估计结果 $\boldsymbol{x}$ 的方法就是 Gauss-Seidel 迭代。不像 Jacobi 方法等到下一次迭代,Gauss-Seidel 迭代可以立即使用每个新的分量进行迭代。

Gauss-Seidel 迭代是

$$x_n^{(k+1)} = (b_n - \sum_{p<n} A_{n,p} x_p^{(k+1)} - \sum_{p>n} A_{n,p} x_n^{(k)}) / A_{n,n} \tag{4.17}$$

Gauss-Seidel 立即把每个新元素 $x_n^{(k+1)}$ 替换旧的 $x_n^{(k)}$，从而节省了存储空间，使得编程更加简单。另一方面，Gauss-Seidel 迭代只能依次执行，因为 $\boldsymbol{x}^{(k+1)}$ 的每个分量只依赖于先前的数值；Gauss-Seidel 本质上是顺序执行的。Gauss-Seidel 也需要一些条件来保证其收敛性；这些条件比 Jacobi 要宽松（比如矩阵是对称正定的）。Gauss-Seidel 方法收敛速度快，仅需要 Jacobi 方法稍多一半的迭代次数就可以获得相同的精度。根据式(4.17)，Gauss-Seidel 方法也要求矩阵 $\boldsymbol{A}$ 满足对角线元素非零。Gauss-Seidel 迭代被广泛使用，如自适应滤波。

松弛是通过修改一个未知参数来修正方程的过程。Jacobi 是并行松弛，Gauss-Seidel 是连续松弛。超松驰是一种较大修正的技术，而不仅是严格满足等式方程的修正。超松弛可以大大加速收敛速度。逐次超松弛法简称 SOR 法，是 Gauss-Seidel 法的加速法；是把 Gauss-Seidel 之前的迭代结果和当前结果进行加权平均，即

$$\boldsymbol{x}^{(k+1)} = \omega \, \boldsymbol{x}_{GS}^{(k+1)} + (1-\omega) \, \boldsymbol{x}^{(k)} \tag{4.18}$$

其中 $\boldsymbol{x}_{GS}^{(k+1)}$ 是 Gauss-Seidel 下一次迭代结果，$\omega > 1$ 是松弛因子。ω 值决定了收敛速度的加速度。如果 $\omega = 1$，则 SOR 方法就是 Gauss-Seidel 方法。如果 $\omega < 1$，这就是欠松弛，导致收敛缓慢。如果使用最优 ω，SOR 会比 Gauss-Seidel 快几个数量级。然而，除了特殊的矩阵之外，选择最优 ω 一般是困难的。由于 SOR 是逐次松弛（Gauss-Seidel 迭代），所以它只需要保留 $\boldsymbol{x}$ 的一个副本。

与直接法相比，迭代方法有以下几个优点：①比直接法需要更少的内存；②比直接法收敛更快；③可以用更简单的方式处理特殊的矩阵结构（如稀疏矩阵）。此外，迭代技术可以通过一个好的初始值，减少求解所需的迭代次数。尽管在理论上收敛到最优解需要若干次迭代，但迭代技术能够根据所需的精度水平随时停止迭代。而直接法，它们没有利用初始值，只是执行预定的操作序列，并在所有操作之后求得解。

但是，迭代方法运算是复杂的。非定常迭代法例如上面所提到的梯度法和共轭梯度方法，虽然非常高效但是每次迭代的操作复杂度是 $O(N^2)$。定常迭代法，如 Jacobi、Gauss-Seidel 和 SOR 算法，每次迭代复杂度是 $O(N)$，复杂度降低但是效率也变低。迭代法包含除法和乘法操作，使得它们对于实时信号处理和硬件设计来说是昂贵的。

DCD 算法是基于定常坐标下降的非定常迭代法。它的实现很简单，不需要乘法或除法操作。每次迭代，它只需要 $O(N)$ 次加法或 $O(1)$ 次加法。因此，DCD 算法非常适合硬件实现。

4.2 硬件设计

长期以来，在实时系统中求解线性方程组和矩阵求逆是很难实现的部分。最近开始出现相关硬件设计。大多数相关的硬件设计是基于 Givens 旋转的 QRD 技术。

Karkooti 等人在 Xilinx Virtex-4 XC4VLX200 FPGA 芯片上实现了基于 SGR 的 QRD 算法 4 × 4 浮点型复数矩阵求逆。该设计使用 21 位数据格式；其中小数部分 14 位，浮点数指数 6 位，符号位 1 位。该设计是用 Xilinx System Generator tool 实现的，通过 Xilinx Core Generator 实现浮点除法器模块。为了把整个算法应用于单个芯片，设计采用了组合型的脉冲阵列结构。该设计包含一个对角线单元，一个非对角线内部单元和一个向后置换块。该设计还需要块 RAM 和一个控制单元来调度这些 RAM 之间的数据移动。和三角形架构脉冲阵列相比，这种组合结构以较大延迟为代价减少了复杂度。该设计资源约为 9117 个逻辑片和 22 个 DSP48（也称为"XtremeDSP"）模块，延迟时间为 933 个时钟周期（QRD 为 777 个周期，向后置换为 156 个周期）。Virtex-4 的 DSP48 模块是一个基于 18 位×18 位硬件乘法器的可配置乘法累加模块。稍微修改控制单元和 RAM 的大小，可将设计扩展到其他尺寸的矩阵。

Edman 等人在 Xilinx Virtex-II FPGA 芯片上使用 SGR 的 QRD 技术实现了 4×4 定点型复数矩阵求逆。该设计是基于线性结构脉冲阵列，该阵列通过三角形架构脉冲阵列的直接投影获得。这个线性阵列仅需要 $2N$ 个处理单元进行 QRD 和后置置换。但是，这些处理单元非常复杂，而且这些处理单元不都是 100% 的利用率。最复杂的复数除法器需要 9 个乘法器，3 个加法器和一个具有 5 个流水线阶段的查询表。这个设计采用 19 位定点数据格式，占用了 Virtex-II 芯片资源面积的 86%，延迟时间为 175 个周期。然而，文献没有详细指出 FPGA 芯片模型和逻辑片，RAM 和乘法器的资源使用情况。

刘等人在台湾半导体电路制造股份有限公司（TSMC）0.13 微米的芯片上实现了一种用于自适应波束形成的 SGR 浮点型 QRD 阵列。该设计是使用折叠和映射方法获得 N 个单元线性架构脉冲阵列。处理单元与三角形脉冲阵列中的单元具有相似的复杂性，有 100% 的利用率。通过使用参数化算术处理器，提供了用非常规范直接的方法来实现 QR 阵列处理器的通用核。对于数据位是由 14 位尾数和 5 位指数表示的 41 根天线阵列，最大时钟速率是 150 MHz 的线性阵列 QR 处理器包括 21 个处理单元和 1060K 个逻辑门。该文献作者没有采用 FPGA 芯片实现这个线性阵列。由于 SGR 的基本操作是非常复杂的，我们可以得出这样的结论，FPGA 实现这个线性阵列是非常复杂的，并且不可能在单个 FPGA 芯片上实现诸如 41 个元素的大规模系统。

Myllyla 等人在 Xilinx Viretex-II XC2V6000 FPGA 芯片上实现了 2×2 和 4×4 MIMO OFDM 系统的定点型复数最小均方误差检测器。该设计的矩阵运算基于脉冲阵列，用基于 CORDIC 的 QRD 技术和基于 SGR 的 QRD 技术完成。对于 2×2 天线系统，该设计采用了脉冲阵列的快速平行三角形结构，针对 4×4 天线系统，该设计采用了易于伸缩和时间共享处理单元的简单线性体系结构。基于 CORDIC 的设计是用 VHDL 实现的，而基于 SGR 的设计是使用 System Generator 实现的。对于 2×2 和 4×4 系统，CORDIC QRD 设计采用 16 位定点型数据格式，分别需要 11910 和 16805 个逻辑片，6 和 101 个块 RAM，20 和 44 个 18 位×18 位嵌入式乘法器，伴随 685 个和 3000 个周期的延迟。基于 SGR 的 QRD 技术仅适用于 2×2 系统，采用 19 位定点数据格式，它需要 6305 个逻辑片，8 个块 RAM 和 59 个延迟时间为 415 个周期的 18 位×18 位嵌入式乘法器。很明显，与基于 SGR 的设计相比，基于 CORDIC 的设计需要更多的逻辑片和更少的乘法器，这是因为 CORDIC

是免乘法和除法器的旋转操作。

另外,还有许多商用 QR 知识产权(IP)内核使用 CORDIC 算法。Altera 发布了采用 CORDIC IP 内核的 CORDIC-QRD-RLS 设计,使其可以实现智能天线波束成形、WiMAX、3G 无线通信信道估计和均衡等应用。Altera 的 CORDIC IP 模块采用深度流水线并行架构,使得它们在 Stratix FPGA 上的速度超过 250 MHz。文献[36][37]中研究了在 Altera Stratix FPGA 上用 CORDIC 块对 16 位数据位的 9 元素天线的 64 个输入向量进行矩阵分解的不同程度并行性。CORDIC 内核所需的逻辑资源可低至 2600 个逻辑元件(相当于 1300 个 Xilinx 逻辑片[41,42])。在 Altera Stratix FPGA 上当 CORDIC 内核以 150 MHz 运行时,该设计得到一个 5 kHz 的更新速率,处理延迟约为 29700 个周期。而其他模块的逻辑资源不详。嵌入式 NIOS 处理器执行 9 × 9 矩阵的向后替换操作需要约 12000 个周期延迟。然而,QRD 和向后替换不能以流水线方式执行。因此,设计的总延迟约为 41700 个周期。

Xilinx 拥有类似的 CORDIC IP 内核。Dick 等人用 Xilinx System Generator tool,在 Virtex-4 FPGA 器件上实现了向后替换复数型折叠 QRD。折叠设计包含脉冲阵列中的一个对角线单元(基于 CORDIC),一个非对角线单元(基于 DSP48 的),以及一个向后替换单元,另外需要块 RAM 和一个控制单元来调度这些块之间的数据移动。所有这些需要大约 3530 个逻辑片,13 个 DSP48 块和 6 个块 RAM。为了求解一个 9×9 方程组,所提出的设计约需要 10971 个周期。与 Dick 2005 年所提出的基于 SGR 的 QRD 4×4 矩阵求逆核相比,基于 CORDIC 的 QRD 所占用的资源面积要小得多;这是因为它使用定点型而不是浮点型架构,SGR 操作是基于正常的算术运算,而 CORDIC 操作是免乘法器和除法器的旋转操作运算。

Xilinx 还拥有最初源于 AccelChip 公司的 AccelWare DSP IP Toolkits 所提供的 QR 分解,QR 求逆和 QRD-RLS 空间滤波 IP 内核。AccelWare 是一个浮点型 Matlab 模型生成器库,可由 AccelDSP 综合生成高效的定点型硬件。这些 AccelWare DSP 内核可用于 WiMAX 基带 MIMO 系统和波束形成应用。Uribe 等人[12]研究了 4 元素的基于 CORDIC 的 Givens 旋转的 QRD-RLS 算法波束形成器。在器件(Xilinx Virtex-4 XC4VSX55,FPGA)上所需的资源是 3076 个逻辑片和一个 DSP48 块。与文献[44]中所提的设计相比,它所需的逻辑资源和 DSP48 块的数量减少了;这是因为[12]中的设计主要是基于 CORDIC 操作,而文献[44]中的设计仅对角线单元是基于 CORDIC 操作的。该设计的采样吞吐率为 1.7 MHz。他们选择的器件(XC4VSX55)有 400 MHz、450 MHz 和 500 MHz,分别可以提供 235、265 和 294 个周期,作者没有给出他们正在使用设备的时钟速率。

Matousek 等人用高速对数算术(HSLA)库实现了脉冲阵列的对角线单元的 QRD。设计采用了 LNS 格式,例如数字总是 8 位的整数部分和取决于数据精度大小的小数部分。对于 32 位 LNS 格式,其 QRD 对角线单元大约需要 3000 个 Xilinx Virtex 逻辑片,延迟时间是 13 个周期。为了进行比较,他们还实现了由 23 位尾数和 8 位指数组成数据的浮点型 QRD 对角线单元,其需要延迟为 84 个周期的 3500 个逻辑片。很明显,基于 LNS 算法的设计比传统基于算术的设计要快得多。

Schier 等人使用与[47]中相同的 HSLA 库在 Xilinx Virtex-E XCV2000E FPGA 上实

现了 Givens 旋转浮点型操作和基于 QRD 的 RLS(QRD-RLS)算法。使用了两种 LNS 数据格式:一种是 19 位,另一种是 32 位。设计仅需要一个对角线单元和一个非对角线单元的脉冲阵列,而不是完整的阵列。由于加法和减法成为 LNS 中计算量最大的复杂模块系统,可以使用一阶泰勒级数查询表近似评估。甚至可以通过错误纠正机制和范围移动算法来最小化查询表大小,对数加法/减法模块仍然需要大量的逻辑片和存储空间。对于 19 位和 32 位 LNS 格式,一个 Xilinx Virtex-E XCV2000E FPGA 的一个加/减模块分别需要约 8% 和 13% 的逻辑资源,以及 3% 和 70% 的块 RAM。因此,对于 19 位 LNS 格式,这两个单元(一个对角单元和一个非对角单元)需要大约 4492 个逻辑片和 30 个块 RAM。对角线单元有 11 个周期延迟,非对角线单元有 10 个周期延迟。这两种单元都是完全流水线方式,在 75 MHz 的工作频率上工作,因此可以获得约为 6.8 MHz 的吞吐量。与之相对应的,32 位 LNS 数据格式的设计,只能在 Xilinx XCV2000E FPGA 上配置一个对数加法/减法模块。

Eilert 等人在 Xilinx Virtex-4 FPGA 上实施了基于 BAMI 算法的浮点型 4 × 4 复数型矩阵求逆核。该设计用 Xilinx Core Generator 生成所有基本单元,如浮点型加法器、减法器、实数乘法器和实数除法器。由于乘法和除法是使用逻辑门执行的,而不是嵌入式乘法器或 DSP48 模块,因此 BAMI 设计需要较大的芯片资源面积。它总共分别需要 16 位和 20 位浮点型数据 7312 个逻辑片和 9474 个逻辑片。16 位设计和 20 位设计分别运行在 120 MHz 和 110 MHz 工作频率上,都伴随 270 个周期延迟。

LaRoche 等人使用 Xilinx Virtex II XC2V600 FPGA 芯片的 Xilinx Synthesis Technology (XST)综合了修正 Sherman-Morrison 方程(4.9)的一个 4 ×4 复数矩阵求逆单元。尽管修正后的 Sherman-Morrison 方程(4.9)不包含除法运算,但由于其包含大量的乘法运算,其 FPGA 实现仍然非常复杂。该设计具有矩阵-矩阵乘法模块,矩阵-向量乘法模块,向量-向量乘法模块和标量-矩阵乘法模块四个主要模块,分别消耗大约 3108、765、187 和 780 个逻辑片和 64、16、4 和 16 个 18 位×18 位嵌入式乘法器。文献[24]给出的逻辑资源是 4446 个逻辑片和 101 个 18 位×18 位嵌入式乘法器,小于四个主要模块的资源使用总和。但是作者没有资源区别的解释。而且,也没有给出所需 RAMs 数量。该设计的延迟约为 64 个周期。

表 4-1 基于 FPGA 的矩阵求逆与线性方程求解器的比较(MULT =乘法器)

矩阵规模	技术	逻辑片	其他	周期数
2×2	QRD-SGR[4]	6305	59 MULTs	415
4×4	QRD-SGR[15]	9117	22 DSP48s	933
2×2	QRD-CORDIC[4]	11910	20 MULTs	685
4×4	QRD-CORDIC[4]	16805	44 MULTs	3000
4×4	QRD-CORDIC[12]	3076	1 DSP48s	265
9×9	QRD-CORDIC[36]	1300	1NIOS Processor	41700

续表 4-1

矩阵规模	技术	逻辑片	其他	周期数
9×9	QRD-CORDIC[44]	3530	13 DSP48s	10971
4×4	BAMI[23]	9474	—	270
4×4	Modified Sherman-Morrison[24]	4446	101 MULTs	64
1 diagonal cell	QRD-LNS[47]	3000	—	13
1 diagonal cell	QRD[47]	3500	—	84
1 diagonal cell 1 off-diagonal cell	QRD-LNS[49][48]	4492	—	11 10

表 4-1 比较了上面提到的算法 FPGA 实现。可以看出,目前解决正则方程和矩阵求逆问题的方法需要相对较高的计算资源,尤其是使用硬件乘法器。在实时硬件设计中,仅小规模的问题能得到有效的解决。

第5章 多用户检测

CDMA 系统中多个用户占用同一时隙和同一频带,使用不同的地址码,从而产生多址干扰问题。在现实系统中,理想的完全同步很难实现,地址码也难以保持完全的正交性,使得多址干扰成为 CDMA 通信系统中的一个主要干扰。传统的检测技术按照经典直接序列扩频理论对用户的信号分别进行扩频码匹配处理,不能从根本上消除多址干扰。多用户检测算法,可以降低系统对功率控制精度的要求,从而显著提高系统频谱效率。多用户干扰是有明显结构性的伪代码序列,如果用户间的相关函数已知,则可以通过已知信息抑制其影响。其中传统单用户检测方法可以采用匹配滤波器,然而它受远近问题影响严重且性能较差。最大似然(ML)检测方法可以提供最优的检测性能,但是它的复杂度太高不适用于实际操作。解相关检测方法虽然不需要知道接收信号幅度,但是它涉及矩阵求逆运算量较大而且解相关操作增强了背景高斯噪声功率。最小均方误差检测方法考虑了背景白噪声影响对相关矩阵进行修正取逆从而抑制多址干扰,然而它同样需要涉及矩阵求逆。判决反馈(DF)结合串行干扰消除(或者并行干扰消除)可以明显减少多址干扰,但是如果初始判决不可靠将对下级产生较大干扰。球形约束和盒型约束算法是把检测结果限制在一个封闭的凸面集合里面,该约束可以降低检测错误。半正定松弛方法把 ML 问题转换成一个半正定问题,其检测性能非常接近 ML 算法,但是针对大规模用户系统,半正定松弛方法运算量巨大。球形译码和分支定界算法可以取得最优检测性能,但是它们对于"最差情况"的检测运算量非常高。

5.1 多用户检测系统模型

5.1.1 同步 CDMA 系统

在同步 CDMA 系统中,接收到信号是:

$$y(t)=\sum_{k=1}^{K}A_k x_k s_k(t)+n(t) \tag{5.1}$$

其中

- $s_k(t)$ 是第 k 个用户的特征波形。特征波形 $s_k(t)$ 可以表示为

$$s_k(t)=[p(t),\ p(t-T_c),\ p(t-(m-1)T_c)]s_k \tag{5.2}$$

其中 $p(t)$ 是码片波形,T_c 是码片周期,s_k 是 $m\times 1$ 向量并称为用户 k 的扩频序列;m

是扩频因子。

· x_k 是第 k 个用户的输入信息符号。

· A_k 是第 k 个接收信息符号的幅度。

· $n(t)$ 是功率谱密度是 σ^2 的零均值加性白高斯噪声。

特征波形的相关性定义为

$$\rho_{ij} = \int_0^{T_s} s_i(t) s_j(t) \mathrm{d}t \tag{5.3}$$

其中 T_s 是符号周期。我们将相关矩阵定义为: $\boldsymbol{R} = \{\rho_{ij}\}$;$\boldsymbol{R}$ 是 $K \times K$ 的对称非负正定矩阵。

5.1.2 异步 CDMA 系统

在异步传输的情况下,接收信号可以表示为:

$$y(t) = \sum_{k=1}^{K} \sum_{i=0}^{N} A_k x_k[i] s_k(t - iT_s - \tau_k) + n(t) \tag{5.4}$$

其中 t 是传输时间,τ_k 是第 k 个用户的传输时延,$x_k[i]$ 是第 k 个用户的第 i 个发送符号,$(N+1)$ 是总数据块长度。用户发送符号流: $x_k[0],\cdots,x_k[N]$ 。如果信道状态信息是已知的,也就是说接收机知道所有时延 τ_k 和信道增益 A_k ,我们可以把异步 CDMA 系统看作是用户数为 $K(N+1)$ 的同步 CDMA 的情况。扩频序列修改后是:

$$\tilde{s}_{k,i}(t) = s_k(t - iT_s - \tau_k) \tag{5.5}$$

修改后的信道增益为 $\tilde{A}_{k,i} = A_k$,修改后的数据符号为 $\tilde{x}_{k,i} = x_k[i]$, $k = 1,\cdots,K$ 和 $i = 0,\cdots,N$ 。修改后的信号模型可以表示为:

$$y(t) = \sum_{k=1}^{K(N+1)} \tilde{A}_k \tilde{x}_k \tilde{s}_k(t) + n(t) \tag{5.6}$$

因此,当用户数量从 K 增加到 $K(N+1)$ 时,异步 CDMA 系统可以看作是(5.1)同步 CDMA 系统模型的特殊情况。

5.1.3 平坦性衰落 CDMA 系统

频率平坦性缓慢衰落会影响接收信号幅度,但不会引起特征波形失真。因此,同步 CDMA 系统模型(5.1)的数学公式也适用于描述缓慢频率平坦性衰落信道。快速频率平坦性衰落会影响接收信号幅度并会导致特征波形失真。如果把特征波形修改为:

$$\tilde{s}_k(t) = A_k(t) s_k(t) \tag{5.7}$$

那么同步 CDMA 模型(5.1)也可以适用于快速频率平坦性衰落信道:

$$y(t) = \sum_{k=1}^{K} x_k \tilde{s}_k(t) + n(t), \qquad t \in [0, T_s]. \tag{5.8}$$

5.1.4 频率选择性衰落 CDMA 系统

在频率选择性快速衰落信道中，第 k 个用户的特征波形经过脉冲响应 $r_k(\tau,t)$ 。频率选择性衰落信道中的变换特征波形 $\tilde{s}_k(t)$ 可表示为：

$$\tilde{s}_k(t) = \int_{-\infty}^{\infty} r_k(\tau,t)s_k(t-\tau)\mathrm{d}\tau \tag{5.9}$$

依然可以使用同步 CDMA 模型：

$$y(t) = \sum_{k=1}^{K} x_k \tilde{s}_k(t) + n(t) \tag{5.10}$$

等式(5.10)表明频率选择性衰落信道模型可以转化为同步 CDMA 系统模型。

5.2 多用户检测器

多址干扰严重限制了传输系统的性能和容量。多用户检测技术可以减少这种干扰，并缓解远近距离问题。最大似然多用户检测器可以提供接近单用户检测的最优检测性能，但它的复杂度与用户数呈指数关系增长。解相关检测器可以消除多址干扰，但同时增强了噪声功率。最小均方误差检测器比解相关检测器的性能好，但它需要估计幅值和矩阵求逆操作。因为操作简单并提供良好的性能，判决反馈(DF)检测器成为当下流行的检测法之一。球型约束和盒型约束算法把结果限制在一个封闭的凸集里，显著提高了性能。半正定松弛算法将最大似然问题转化成一个半正定问题，并提供与最大似然检测器非常接近的误码率性能。概率数据关联(PDA)检测器把多址干扰看成有匹配均值和协方差的高斯噪声，提供接近最大似然检测器的检测性能。球形译码(SD)和分支定界(BB)检测可以取得最优检测性能，但是它们对于“最差情况”的检测所需运算量非常高。文献[15]的研究表明从复杂度和检测性能的角度来看，DF 检测器、PDA 检测器和 BB 检测器组成了众多高级多用户检测器的“有效边界”。

5.2.1 传统匹配滤波器

匹配滤波器是最传统的检测器，它把接收信号与期望用户的扩频波形卷积，如图5-1所示。

第 k 个匹配滤波器的输出 θ_k 是：

$$\theta_k = \int_0^{T_s} y(t)s_k(t)\mathrm{d}t \tag{5.11}$$

其中 $y(t)$ 是接收信号，也可以写作：

$$\theta_k = \int_0^{T_s} \left\{ \sum_{j=1}^{K} A_j x_j s_j(t) + n(t) \right\} s_k(t)\mathrm{d}t \tag{5.12}$$

利用等式(5.3)到(5.12)，我们可以得到：

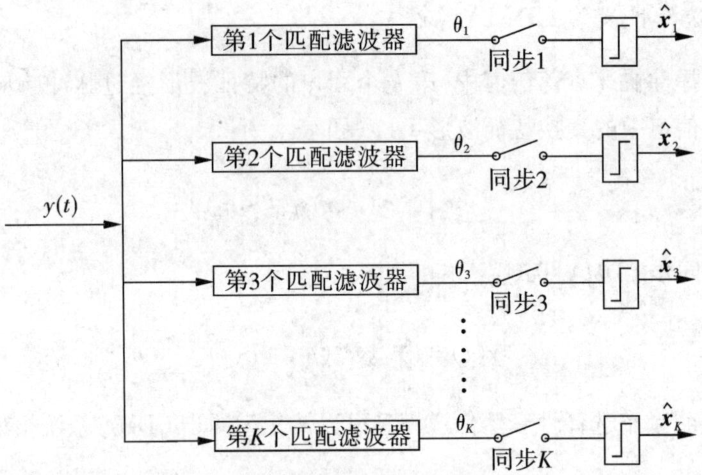

图 5-1　传统匹滤波器

$$\theta_k = \sum_{j=1}^{K} A_j x_j \rho_{jk} + n_k \tag{5.13}$$

其中

$$n_k = \int_0^{T_s} n(t) s_k(t) \mathrm{d}t \tag{5.14}$$

所以可以得到

$$\theta_k = A_k x_k + \sum_{\substack{j=1 \\ j \neq k}}^{K} A_j x_j \rho_{jk} + n_k \tag{5.15}$$

公式(5.15)的第二项是多址干扰。匹配滤波器把多址干扰当作是加性高斯白噪声。但是多址干扰严重影响了匹配滤波器的性能和容量。

当干扰用户数量增加时,多址干扰会增强。此外,能量幅值较大的用户会对能量幅值较低的用户产生更大的干扰。距离发射端近的信号比距离发射端远的信号的幅值衰减小,这就是远近效应问题。传统匹配滤波器只需要知道扩频序列信息。然而,当用户数量增加时,匹配滤波器的检测性能会变差。

5.2.2　最大似然检测器

假设 AWGN 信道中有 K 个用户的同步 CDMA 系统。匹配滤波器的输出是:

$$\boldsymbol{\theta} = \boldsymbol{RAx} + \boldsymbol{n} \tag{5.16}$$

如果信号是 BPSK 调制,向量 $\boldsymbol{x} \in \{-1, +1\}^K$ 包含 K 个用户发送的信息符号,$\boldsymbol{R}$ 是正定扩频序列相关矩阵;$\boldsymbol{A}$ 是对角矩阵,A_{kk} 是它的第 k 个对角元素,也是第 k 个用户接收信号能量的平方根,$\boldsymbol{n}$ 是实数协方差矩阵 $\sigma^2\boldsymbol{R}$ 的零均值高斯随机向量。

最大似然多用户检测器用最小化二次代价函数得到向量 $\boldsymbol{x}$ 的估计值。

$$\hat{\boldsymbol{x}} = \arg \min_{\boldsymbol{x} \in \{-1,+1\}^K} J(\boldsymbol{x}) \tag{5.17}$$

其中二次代价函数 $J(\boldsymbol{x})$ 表示为：

$$J(\boldsymbol{x}) = \| \boldsymbol{\theta} - \boldsymbol{R}\boldsymbol{x} \|^2 \Rightarrow \boldsymbol{x}^{\mathrm{T}}\boldsymbol{A}\boldsymbol{R}\boldsymbol{A}\boldsymbol{x} - 2\,\boldsymbol{x}^{\mathrm{T}}\boldsymbol{A}\boldsymbol{\theta} \tag{5.18}$$

最大似然检测算法的复杂度是 $O(2^K)$ 个算术操作数。

虽然最大似然检测器可以提供最佳的检测性能，但它的复杂度随用户数量呈指数增长，使得硬件难以实现。

5.2.3 解相关检测器

在式(5.16)中，传输数据可以写成

$$\hat{\boldsymbol{x}} = \mathrm{sign}(\boldsymbol{R}^{-1}(\boldsymbol{R}\boldsymbol{A}\boldsymbol{x} + \boldsymbol{n})) = \mathrm{sign}(\boldsymbol{A}\boldsymbol{x} + \boldsymbol{R}^{-1}\boldsymbol{n}) \tag{5.19}$$

如果 $\sigma = 0$, $\hat{\boldsymbol{x}} = \mathrm{sign}(\boldsymbol{x})$，那么这种检测方法叫做解相关算法。解相关检测器的优点在于它不需要知道接收信号的能量幅度。但是，当矩阵 $\boldsymbol{R}$ 是病态矩阵时，(5.19)中的 $\boldsymbol{R}^{-1}\boldsymbol{n}$项会导致噪声功率增强，从而使得错误率增加。

5.2.4 最小均方误差检测器

最小均方误差检测器考虑背景噪声影响，结合接收信号功率，在低 SNR 范围内比解相关检测器检测性能更好。最小均方误差检测器最小化真实值和估计值之间的均方误差。解相关检测中的矩阵 $\boldsymbol{R}$ 的倒数被替换为 $[\boldsymbol{R} + \sigma^2 \boldsymbol{A}^{-2}]^{-1}$。

最小均方误差检测器需要在消除多址干扰和噪声增强之间进行权衡。当噪声方差趋于无穷大时，最小均方误差检测器和传统匹配滤波器性能相似。当 SNR 趋于无穷大时，最小均方误差检测器和解相关检测器性能接近。最小均方误差检测器可以很好地解决远近效应问题。

5.2.5 判决反馈检测器

判决反馈检测器一般和串行干扰消除、并行干扰消除、多级或迭代检测器共同使用。串行干扰消除判决反馈检测器(S-DF)可以最大化同步 AWGN 信道容量。S-DF 算法能够减小误差传播，但是它通常导致用户性能不统一。在 S-DF 算法中，用户排序是非常重要的。文献[21]研究了使用最优用户排序的解相关 DF 算法在不完美反馈情况的性能。文献[25]研究了使用最优用户排序的解相关 DF 算法在完美反馈情况的性能。最优排序算法的问题在于，接收机需要非常高的计算负担。并行干扰消除判决反馈检测器(P-DF)接收机可以抵消小区内干扰并抑制其他小区干扰，总体上可以为用户群体提供一致的性能，但是它对于错误传播更为敏感。文献[15,16]提出的多级和迭代判决反馈是串行干扰消除、并行干扰消除和判决反馈的组合，研究表明这种组合比传统 S-DF、P-DF[18]的性能更好。

5.2.6 半正定松弛检测器

松弛是对优化问题的有效逼近技术。放宽一些约束可以简化问题。半正定松弛算法可以在不损失性能的情况下,降低计算复杂度。文献[19-21]描述了用于求解布尔二次规划问题的半正定理松弛算法。

$$\arg\min_{x}(x^{T}Qx) \tag{5.20}$$

其中 Q 是对称矩阵。由于 $x^{T}Qx = \mathrm{Trace}(x x^{T}Q)$,式(5.20)可以重构为:

$$\begin{aligned}&\arg\min_{X}\mathrm{Trace}(QX)\\&\text{s.t. }\ \mathrm{diag}(X)=e\\&X=x x^{T}\end{aligned} \tag{5.21}$$

其中 e 是所有元素为 1 的向量。

约束条件 $X=x x^{T}$ 表明 X 是对称的,半正定的并且秩为 1。式(5.21)是非凸优化问题。如果式(5.21)去除掉 rank-1 的约束时,可以得到:

$$\begin{aligned}&\arg\min_{X}\mathrm{Trace}(QX)\\&\text{s.t. } X\succeq 0\\&X_{jj}=1, j=1,\cdots,K\end{aligned} \tag{5.22}$$

其中 $X\succeq 0$ 表示 X 对称的和半正定的。式(5.22)就是式(5.21)的松弛,因为式(5.21)中的集合是式(5.22)中的子集。式(5.22)是式(5.21)的半正定性松弛。式(5.22)的复杂度是 $O(K^{3.5})$ 。

如果用半正定松弛算法解决最大似然问题,最大似然检测必须构造成和(5.20)形式相同。假设标量 $c\in\{-1,+1\}$ 。由于 $cx\in\{-1,+1\}^{K}$ 属于 $x\in\{-1,+1\}^{K}$,因此式(5.17)可以重写为:

$$\begin{aligned}\max_{x\in\{-1,+1\}^{K}} J(x) &\equiv \max_{\substack{x\in\{-1,+1\}^{K}\\ c\in\{-1,+1\}}} J(cx)\\&=\max_{\substack{x\in\{-1,+1\}^{K}\\ c\in\{-1,+1\}}} 2c\, x^{T}Ay - x^{T}ARAx\\&=\max_{\substack{x\in\{-1,+1\}^{K}\\ c\in\{-1,+1\}}} [x^{T}\ \ c]\begin{bmatrix}-ARA & Ay\\(Ay)^{T} & 0\end{bmatrix}\begin{bmatrix}X\\c\end{bmatrix}\end{aligned} \tag{5.23}$$

这个等效于(5.20)中的布尔二次规划问题

$$Q=\begin{bmatrix}-ARA & Ay\\(Ay)^{T} & 0\end{bmatrix}$$

在用户之间的相关性较强或远近效应明显的情况下,半正定松弛检测器可以提供接近最大似然检测器的 BER 性能。然而,半正定松弛检测器对于大规模系统来说是非常复杂的。

5.2.7 约束性多用户检测器

最大似然检测器可以在 $x\in\{-1,+1\}^{K}$ 的集合中找到 BPSK 调制信号的解,其中

$\{-1, +1\}^K$ 表示所有二进制符号的集合,每个符号是+1 或-1。然而,ML 检测器对实际应用来说太复杂。如果估计向量被限制在封闭凸面集合中,可以降低复杂度。把估计值限制在一个超立方体内就是求解盒型约束二次问题(例如 $K=2$),如图 5-2(a)所示。在 CDMA 系统中,如果用盒型约束来解决最大似然 BPSK 信号,问题被重新转化为:

$$\hat{\boldsymbol{x}} = \arg \min_{\boldsymbol{x} \in [-1,+1]^K} J(\boldsymbol{x}) \tag{5.24}$$

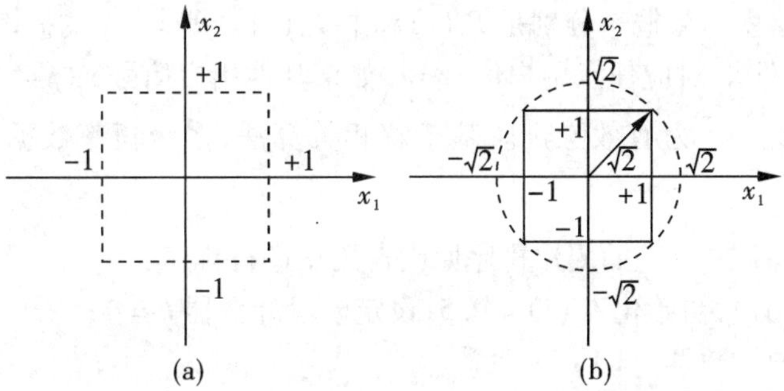

图 5-2 投影到盒型区域(a)和投影在球面区域上(b)

我们把在超立方体封闭凸面集合 $\beta(\beta=[-1,+1])$ 上的正交投影操作叫作 P_B,表示为:

$$P_B(\hat{\boldsymbol{x}}) = \arg \min_{\boldsymbol{x} \in \beta} \| \boldsymbol{x} - \hat{\boldsymbol{x}} \| \tag{5.25}$$

满足:

$$\begin{cases} x_k & \text{if} \quad -1 < x_k < 1 \\ -1 & \text{if} \quad x_k \leqslant -1 \\ +1 & \text{if} \quad x_k \geqslant +1 \end{cases} \tag{5.26}$$

图 5-2(b)是描述球面约束最大似然问题:

$$\hat{\boldsymbol{x}} = \arg \min_{\hat{\boldsymbol{x}} \in S} J(\boldsymbol{x}) \tag{5.27}$$

其中 $S = \{\boldsymbol{x} \in \Re^K : \|\boldsymbol{x}\|^2 \leqslant K\}$ 。假设 $P_S(x_k)$ 是把第 k 个元素正交投影到球面上,表示为:

$$\begin{cases} \alpha x_k & \text{if} \quad \|\boldsymbol{x}\|^2 > K \\ x_k & \text{if} \quad \|\boldsymbol{x}\|^2 \leqslant K \end{cases} \tag{5.28}$$

其中 $\alpha = \sqrt{K} / \|\boldsymbol{x}\|$ 且 $0<\alpha\leqslant 1$。球面约束型多用户检测器和 MMSE 检测器检测性能相似。此外,图 5-2(b)表明球面区域里包含盒型区域[-1,+1],这意味着盒约束检测器具有更严格的约束条件,比球面约束检测器性能更好。

5.2.8 概率数据关联检测器

如果把式(5.16)两边同乘 $\boldsymbol{A}^{-1}\boldsymbol{R}^{-1}$，式子可以表示为：

$$\bar{\boldsymbol{\theta}} = \boldsymbol{x} + \bar{\boldsymbol{n}} = x_i\boldsymbol{e}_i + \sum_{j\neq i} x_j\boldsymbol{e}_j + \bar{\boldsymbol{n}} \tag{5.29}$$

其中 $\bar{\boldsymbol{\theta}} = \boldsymbol{A}^{-1}\boldsymbol{R}^{-1}\boldsymbol{\theta}, \bar{\boldsymbol{n}} = \boldsymbol{A}^{-1}\boldsymbol{R}^{-1}\boldsymbol{n}$，$\boldsymbol{e}_i$ 是第 i 个元素是1而其他元素为零的列向量。公式(5.29)实际上是在硬判决之前的解相关器输出的归一化版本。用户 i 传输信息可以是+1或者-1，出现这些结果概率分别是 $P_x(i)$ 和 $1-P_x(i)$。即 $x_i=1$ 的概率是 $P_x(i)$，$x_i=-1$ 的概率是 $1-P_x(i)$。对应任意用户信号 x_i，如果其他用户信号 $x_j(j\neq i)$ 是二进制随机变量，那么 $\sum_{j\neq i} x_j\boldsymbol{e}_j + \bar{\boldsymbol{n}}$ 为有效噪声。基于解相关算法，多级概率数据关联算法的描述如下：

1. 根据文献[13]中DF的用户排序原则对用户进行排序；
2. 对于所有用户，初始化 $P_x(i)=0.5$，设定阶段计数器 $k=1$；
3. 初始化用户计数器 $i=1$；
4. 根据用户 i 的当前 $P_x(j)(j\neq i)$，更新 $P_x(i)$：

$$P_x(i) = P\left\{x_i = 1 \,\middle|\, \bar{\boldsymbol{\theta}}, \{P_x(j)\}_{j\neq i}\right\} \tag{5.30}$$

5. 如果 $i<K$，则令 $i=i+1$ 并转到步骤4；
6. 如果 $\forall i, P_x(i)$ 已收敛，转到步骤7。否则，令 $k=k+1$ 并转到步骤3；
7. 通过下面的方式判决用户 i 的信号 x_i

$$\begin{cases}1 & P_x(i)\geqslant 0.5\\ -1 & P_x(i)<0.5\end{cases} \tag{5.31}$$

计算(5.31)的运算量随用户数量的增加呈指数型增长，定义

$$\breve{\boldsymbol{n}}_i = \sum_{j\neq i} x_j\boldsymbol{e}_j + \bar{\boldsymbol{n}} \tag{5.32}$$

$\breve{\boldsymbol{n}}_i$ 的平均值和协方差矩阵分别是：

$$\begin{aligned}E(\breve{n}_i) &= \sum_{j\neq i} e_j(2P_x(j)-1)\\ \mathrm{Cov}(\breve{n}_i) &= \sum_{j\neq i}[4P_x(j)(1-P_h(j))\boldsymbol{e}_j\boldsymbol{e}_j^{\mathrm{T}}] + \sigma^2\boldsymbol{R}^{-1}\end{aligned} \tag{5.33}$$

定义 $\boldsymbol{\Phi}_i = E(\breve{\boldsymbol{n}}_i)$ 和 $\boldsymbol{\Psi}_i = \mathrm{Cov}(\breve{\boldsymbol{n}}_i)$。更新概率 $P_x(i)$ 是：

$$\frac{P_x(i)}{1-P_x(i)} = \exp\{-2\boldsymbol{\Phi}_i^{\mathrm{T}}\boldsymbol{\Psi}_i^{-1}\boldsymbol{e}_i\} \tag{5.34}$$

直到所有概率 $\{P_{x_i}\}$ 收敛，算法结束。该算法的检测性能在高SNR下接近单用户性能。更多细节参考文献[6]。PDA的复杂度是 $O(K^3)$。

5.2.9　二分坐标下降算法(DCD)

在许多通信系统中,信号检测就是求解 $\boldsymbol{Rx}=\boldsymbol{\theta}$,其中 $\boldsymbol{R}$ 是 $K\times K$ 对称正定矩阵,$\boldsymbol{x}$ 和 $\boldsymbol{\theta}$ 是 $K\times 1$ 的向量。矩阵 $\boldsymbol{R}$ 和向量 $\boldsymbol{\theta}$ 是已知的,解 $\boldsymbol{x}$ 是未知的。

DCD 算法旨在没有乘法和除法操作的情况下为求解矢量 $\boldsymbol{x}$ 提供一个简单的解决方案。解向量 $\boldsymbol{x}$ 的精度取决于位数(M_b),它用来表示幅度在 $[-H,H]$ 范围里的矢量 $\boldsymbol{x}$ 的比特数。算法中的第一组迭代使用步长参数 d 来确定 $\boldsymbol{x}$ 的最重要的有效位($m=1$)。随后的迭代集合用来确定适当精度(最大到 M_b)的较低有效位。$\boldsymbol{r}=\boldsymbol{\theta}-\boldsymbol{R}\bar{\boldsymbol{x}}$ 是初始残差矢量,其中 $\bar{\boldsymbol{x}}$ 是 $\boldsymbol{x}$ 的初始化。表5-1 描述了 DCD 算法。我们分别用 $x(j)$ 和 $r(j)$ 表示向量 $\boldsymbol{x}$ 和 $\boldsymbol{r}$ 的元素。在向量 $\boldsymbol{x}$ 设置为零的情况下, $\boldsymbol{r}$ 等于 $\boldsymbol{\theta}$。步长设为 H,成功的迭代计数器 p 设置为0。步长 d 通过步骤1中2的指数减少,它不需要乘法或除法,所有乘法和除法都可以用简单的位移代替。如果在一次迭代中解向量被更新,则这样的迭代被标记为“成功”。对于每次步长更新,算法重复成功的迭代,直到残差向量 $\boldsymbol{r}$ 里所有元素变得非常小,使得所有 j 元素不满足步骤4的条件。算法的运算量主要取决于成功的更新迭代次数计数器 p 和比特数 M_b 。预先设定的成功迭代次数 N_u 可以作为算法停止的条件。如果没有这样预设,或者预设数值非常高,则解向量的精度是 2^{-M_b+1} 。DCD 算法解决特定系统数学方程的运算量取决于许多因素。然而,DCD 在给定的 N_u 和 M_b ,最坏情况的复杂度是 $K(2N_u+M_b)$ 个移位累积(*SAC*)操作数。

表5-1　二分法坐标下降算法

初始化: $\boldsymbol{x}=\bar{\boldsymbol{x}}$, $\boldsymbol{r}=\boldsymbol{\theta}-\boldsymbol{R}\bar{\boldsymbol{x}}$, $d=H$, $p=0$
for $m=1:M_b$ 1. $d=d/2$ 2. **Flag**=0　　　　…pass 3. **for** $j=1:K$　　　　…iteration 4. **if** $\lvert r(j)\rvert>(d/2)R(j,j)$ **then** 5. $x(j)=x(j)+\mathrm{sign}(r(j))d$ 6. $\boldsymbol{r}=\boldsymbol{r}-\mathrm{sign}(r(j))\cdot d\cdot\boldsymbol{R}(:,j)$ 7. **Flag**=1, $p=p+1$ 8. **if** $p>N_u$ **end algorithm** 9. **end** j**-loop** 10. **if Flag**=1, goto 2 **end** m-loop

5.2.10 格基检测:分支界定检测器

格基检测问题可以追溯到20世纪90年代初。它是解决整数规划问题的最短/最近格基向量问题的理论和算法。最近(格基)向量问题(CVP)(也称为最近的格点问题)属于近邻搜索或最近点查搜索,它的解集是由基格中的所有点组成。文献[31]推导出求解CVP的根格基算法,根格基来自李群代数。这些算法对于实现低复杂度的格基量化器和高斯信道的编码方案都是很重要的。从格基角度来看,*ML*解码可以解决格基中最近向量的问题。然而,最优的多用户检测是*NP-hard*问题,然而有一些次优算法可以用多项式时间的运算量来完成多用户检测。BB算法用分支攻克结构解决组合优化问题。它的主要思想是将离散优化问题的解集分解成一系列的子集(分支),利用每个子集上的成本函数值和限定边界值来去除一些子集。当搜索完整个解集合时,算法停止。因为BB算法可以有效地搜索整个解空间,因此它得到的最佳结果是全局优化。在分支过程中不计算子集,尽早地删除许多子集,是BB算法的关键。罗等人提出了深度优先BB检测算法,证明球解码器是深度优先BB算法的一种。BB算法是树搜索算法,树中的节点表示一个子集,树的根表示整个解空间。每个节点和获取全局最优值的成本下限是相关联的。因此,如果一个节点上的成本超过当前成本,则去除该节点,那么该节点下的子节点不需要进行计算就可以一并去除。该算法保留要处理的节点列表。当一个节点被保留时,它的子分支会被保留,就需要计算它们的成本。只有低于当前成本的子节点会被添加到列表中。但是,在最坏的情况下,这种算法可能不得不保留整棵树。那么它所需要的运算量随搜索树的层数呈指数增长。

表5-2总结和比较了多用户检测算法。这个表中列出了这些检测器的优点、复杂性和缺点。最优最大似然检测器的复杂度$O(2^K)$过高。传统的匹配滤波检测器主要针对单用户,复杂度$O(K)$低。然而,它不考虑系统中的其他用户影响,所以不能提供良好的性能。解相关检测器本质上是传统检测器的输出乘以用户扩频序列相关矩阵的逆矩阵。解相关检测器的复杂度为$O(K^3)$。该检测器不需要知道每个用户估计值或控制的功率,然而由于$\boldsymbol{R}^{-1}\boldsymbol{n}$操作而噪声被增强。MMSE检测器在存在信道噪声的情况下最小化误差的平方,在低SNR中比解相关器性能更好。然而,它需要求逆操作,求逆在FPGA实现是非常复杂的。判决反馈检测器是最流行的多用户检测方法之一,与线性检测器相比,它操作简单并且性能出色。但是,它的性能主要取决于检测顺序。迫零判决反馈检测器需要Cholesky分解和矩阵求逆,这是硬件难以实现的。半正定松弛检测器是精确最大似然检测器的替代方案,但在大规模系统中它的复杂度仍然非常高。盒型约束检测算法对应非线性串行和平行干扰消除的结构。球约束最大似然检测器与MMSE检测器的关系密切。PDA接近单用户检测器的性能,但它需要矩阵求逆来获得噪声的协方差。二分坐标下降算法不需要乘除法,因此适用于硬件实现,然而其性能不如PDA好。BB算法在高SNR的情况下具有较低的平均复杂度,然而在最坏情况下,它的计算复杂度与最佳多用户检测器的计算复杂度相同,即随着K的增加而呈现指数增长。

表 5-2 多用户检测算法比较

名称	复杂度	优点	缺点
最优最大似然检测算法	$O(2^K)$	最优性能	复杂度随系统规模指数性增长
匹配滤波器	$O(K)$	简单	性能差
解相关算法	$O(K^3)$	不需要知道干扰的能量大小信息	需要矩阵求逆
最小均方误差估计算法	$O(K^3)$	在低 SNR 环境下比解相关算法性能较好。	需要矩阵求逆,用户能量幅度和噪声方差
解相关判决算法	$O(K^2)$(未包含矩阵求逆复杂度)	性能优于线性检测算法	性能依赖于检测顺序。需要 Cholesky 分解和矩阵求逆。
半正定松弛	$O(K^{3.6})$	接近 ML 算法性能	对于大规模系统复杂度过高
球形约束算法	$O(K^3)$	接近 MMSE 算法的性能	不能提供最优性能
盒型约束检测算法	$O(K^3)$	性能接近软干扰抵消,优于球形约束检测算法	不能提供最优性能
概率数据关联算法	$O(K^3)$	接近单个用户检测性能	需要矩阵求逆
分支界定	最差情况的复杂度呈 K 指数增长	低平均复杂度	低 SNR 下最差情况的复杂度非常高
二分坐标下降	$K(2N_u + M_b)$	免乘法和除法操作,易于 FPGA 实施	不能提供最优性能

第 6 章 MIMO 系统

本章主要介绍 MIMO 系统模型和 MIMO 系统传输中使用的两种技术：空时编码和空间多路复用。研究表明空时编码结合向前差错纠正编码——卷积码，可以最大化分集数目；空间多路复用可以最大化系统容量。硬件实施是目前面临的主要问题，因为当前算法的硬件实施设计仅可以用于小规模系统。所以高吞吐量和低成本的算法设计是我们未来工作努力的方向。

6.1 基本概念

传统无线传输系统通常是受信道容量限制的单输入单输出（SISO）天线系统。系统无论选择哪种调制方案，始终受到单根天线无线传输的物理限制。为了增加信道容量，需要建立更多的基站，增加传输功率或带宽。因此，可以在发射机配置一根天线的情况下，在接收机配置一个天线阵列或者多根独立的天线来增加接收分集。这就是单输入多输出系统。也可以在发射机上配置一个天线阵列或多根独立的天线，而接收机只配置一根天线，从而降低接收机的复杂度。这种类型就是多输入单输出系统。

此外，当发射机和接收机都配置一个天线阵列或者多根独立天线时，叫作多输入多输出（MIMO）系统。MIMO 技术不仅提供更高的频谱效率或更高的数据速率（高数据速率表明每秒钟在一赫兹带宽上传输更多的比特数），还具有可靠和多样性的优点。尽管 MIMO 系统比较复杂，但它仍然是现代无线通信标准，如 4G，全球微波接入互操作性（WiMAX）和 IEEE 802.11n 的一个重要组成部分。图 6-1 表示四种类型的天线系统模型。

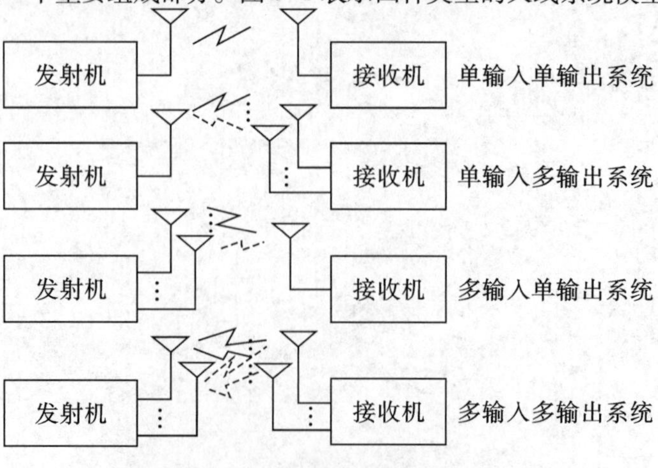

图 6-1 天线系统模型

6.2 MIMO 系统模型

分别在发射端和接收端配置多根天线可以明显提高系统容量。MIMO 系统示意图如图 6-2 所示。

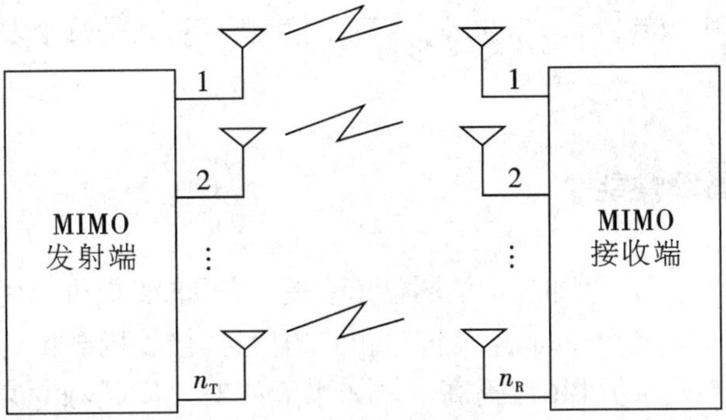

图 6-2 MIMO 系统示意图

空间复用(SM)是把一连串数据首先被分解成多个数据子流,然后数据子流在不同的预处理(编码,调制,时延等)之后从不同的天线被发送出去。每根天线同时发送数据分流。然而,接收端需要有一种合适的算法来分离多根天线上接收到的信号,并恢复原始发送的数据流,从而最大化发送速率和容量。

空时编码(STC)是同时产生和发射天线数量相同的符号并发送出去,一根天线发送一个符号,这些符号是空时编码器生成的,或者是特别设计的在空间和时间上正交传输的符号,接收机用合适的信号处理方法和解码,可以最大化系统地分集增益或编码增益。

我们考虑 n_T 根发射天线和 n_R 根接收天线的 MIMO 系统。$\boldsymbol{x} = [x_1, x_2, \cdots, x_{n_T}]^T$ 是在一个传输周期内每根发射天线发送的数据,$\boldsymbol{r} = [r_1, r_2, \cdots, r_{n_R}]^T$ 是接收数据。接收到的信号可以写成:

$$\boldsymbol{r} = \boldsymbol{H}\boldsymbol{x} + \boldsymbol{n} \tag{6.1}$$

其中 $\boldsymbol{H}$ 是信道矩阵,可以表示为

$$\boldsymbol{H} = \begin{bmatrix} h_{11} & h_{12} & \cdots & h_{1n_T} \\ h_{21} & h_{22} & \cdots & h_{2n_T} \\ \cdots & \cdots & & \cdots \\ h_{n_R1} & h_{n_R2} & \cdots & h_{n_Rn_T} \end{bmatrix}, \tag{6.2}$$

其中 $h_{ij}(i = 1,2,\cdots,n_R; j = 1,2,\cdots,n_T)$ 是发送天线 j 和接收天线 i 之间的信道衰落因子。假设使用独立的瑞利信道模型,h_{ij} 是一个复数随机高斯变量,形式如下:

$$h_{ij} = x + y\iota \tag{6.3}$$

其中 x 和 y 是零均值,方差是 0.5 的独立实数随机高斯数,ι 表示-1 的平方根。

一般来说,我们假定接收总能量与发射总能量是相等的,并进行了归一化,这些能量在信道矩阵里是被平均分配的,如下所示:

$$\sum_{i=1}^{n_R}\sum_{j=1}^{n_T}\left|h_{ij}\right|^2 = n_T \tag{6.4}$$

在公式(6.1)中,$\boldsymbol{n}=[n_1,n_2,\cdots,n_{n_R}]^T$ 表示信道噪声,它是每个发送接收信道上能量为 σ^2 的 AWGN 噪声。

6.3 分集和误码率性能

一般来说,多天线可以增加分集消弱信道衰落。例如,我们可以用单根发射天线和 n 根接收天线。通过 $n_R n_T$ 个不同的路径传输相同的信息,接收机会收到多个独立发送信号衰落副本。因此,接收机的可靠性提高了。如果天线对之间是独立衰落,则可以获得最大的分集增益 n:在高 SNR 下,平均误差概率符合 $1/\mathrm{SNR}^{n_R n_T}$ 曲线,而单根天线衰落系统的平均误差概率是 SNR^{-1}。最近很多工作在研究多根发射天线来获得分集增益。在 n_T 根发射天线和 n_R 根接收天线的系统中,如果每对天线对之间信道是独立衰落的,最大分集数为 $n_R n_T$,它等于发送端和接收端之间独立衰落路径的数量。

如果信号是 BPSK 调制,且信道是瑞利衰落,那么 MIMO 系统的误码率 P_b 和分集数 $n_R n_T$ 的关系是:

$$P_b(\gamma_b)=Q(\sqrt{2\gamma_b}) \tag{6.5}$$

其中每比特的信噪比 γ_b 是:

$$\gamma_b=\frac{\xi}{N_0}\sum_{j=1}^{n_R}\sum_{i=1}^{n_T}h_{ij}^2 \tag{6.6}$$

其中 $\frac{\xi}{N_0}\sum_{j=1}^{n_R}\sum_{i=1}^{n_T}h_{ij}^2$ 是第 i,j 信道上的瞬时 SNR。概率密度函数(PDF) $p(\gamma_b)$ 是

$$p(\gamma_b)=\frac{1}{(n_R n_T)!\ \gamma_c^{-n_R n_T}}\gamma_b^{n_R n_T-1}e^{-\gamma_b/\bar{\gamma}_c} \tag{6.7}$$

假定所有信道都是相同的,那么 $\bar{\gamma}_c$ 是每个信道的平均 SNR。积分得到 P_b:

$$P_b=\int_0^{\infty}P_b(\gamma_b)P(\gamma_b)\mathrm{d}\gamma_b \tag{6.8}$$

公式(6.7)是一个封闭解,可以写作:

$$P_b=[\frac{1}{2}(1-\mu)]^{n_R n_T}\sum_{k=0}^{n_R n_T-1}\binom{n_R n_T-1+K}{k}\left[\frac{1}{2}(1-\mu)\right]^k \tag{6.9}$$

定义

$$\mu = \sqrt{\frac{\bar{\gamma}_c}{1 + \bar{\gamma}_c}} \tag{6.10}$$

当 $\bar{\gamma}_c \gg 1$ 时，$\frac{1}{2}(1+\mu) \approx 1$ 和 $\frac{1}{2}(1-\mu) \approx \frac{1}{4}\bar{\gamma}_c$，并且

$$\sum_{k=1}^{n_R n_T} \binom{n_R n_T - 1 + K}{k} = \binom{2n_R n_T - 1}{n_R n_T} \tag{6.11}$$

所以，当 $\bar{\gamma}_c$ 足够大(大于 10 dB)时，公式(6.9)中的错误概率可以近似为：

$$P_b \approx \left(\frac{1}{4\bar{\gamma}_c}\right)^{n_R n_T} \binom{2n_R n_T - 1}{n_R n_T} \tag{6.12}$$

我们观察到错误概率由 $1/\bar{\gamma}_c$ 增加到 $n_T n_R$ 次方。因此，通过天线分集，错误率与 SNR 的 $n_T n_R$ 次方成反比下降。因此，通过观察高 SNR 情况下的 BER 曲线斜率，就可以得到 MIMO 系统的分集增益。公式(6.9)可以协助评估 MIMO 系统的分集增益。

6.4　空时编码

空时编码是可以最大化 MIMO 系统分集增益的发射分集技术。当有 n_T 根发送天线和一根接收天线时，空时编码可以实现高达 n_T 的分集增益，因此当仅有一根接收天线时，多根发射天线可以降低信道衰落的影响。

主要有两种类型的空时编码：空时网格编码(STTC)和空时分组编码(STBC)。前者类似网格编码调制，而后者实际上是一种映射方案，而不是真正的编码。我们将主要介绍空时分组编码，因为它的实现更简单，并且 STBC 的简易版已纳入 UMTS 标准。

STBC 源于正交设计的数学概念。通过正交设计，所有行和列里的符号都是相互正交的。矩阵里的每一列表示不同的发射天线，行表示不同的发射周期。无线系统需要复数正交设计，也就是说，传输信号可以是符合相应调制的复数。

空时分组编码最简单的形式是“Alamouti 方案”，其示意图如图 6-3 所示。

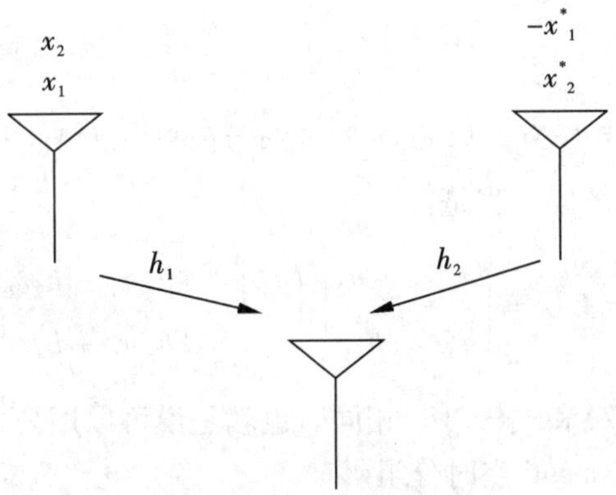

图 6-3　Alamouti 方案示意图

正交编码设计是：

$$\begin{bmatrix} x_1 & x_2^* \\ x_2 & -x_1^* \end{bmatrix} \tag{6.13}$$

在两个连续的周期里，一根天线分别发送 x_1 和 x_2^*，另外一根天线分别发送 x_2 和 $-x_1^*$。在 Alamouti 方案里，解码技术非常简单。根据图 6-3，接收信号是：

$$\begin{aligned} r_1 &= h_1x_1 + h_2x_2 + n_1 \\ r_2 &= h_1x_2^* - h_2x_1^* + n_2 \end{aligned} \tag{6.14}$$

接收机通过线性处理和文献[5]中的 Tarokh 推导，最小化下列式子解码 x_1 和 x_2：

$$|[r_1h_1^*] - (r_2)^*h_2 - x_1|^2 + (-1 + |h_1 + h_2|^2)|x_1|^2 \tag{6.15}$$

$$|[r_1h_1^*] + (r_2)^*h_1 - x_2|^2 + (-1 + |h_1 + h_2|^2)|x_2|^2 \tag{6.16}$$

如果发送符号有恒定的功率，我们可以通过公式(6.15)和(6.16)估计 x_1 和 x_2：

$$\hat{x}_1 = r_1h_1^* - (r_2)^*h_2 \tag{6.17}$$

$$\hat{x}_2 = r_1h_2^* - (r_2)^*h_1 \tag{6.18}$$

式(6.14)与式(6.18)也可以写成：

$$\boldsymbol{r} = \boldsymbol{C}\boldsymbol{x} + \boldsymbol{n} \tag{6.19}$$

信道矩阵 $\boldsymbol{C}$：

$$\boldsymbol{C} = \begin{bmatrix} h_1 & h_2 \\ -h_2^* & h_1^* \end{bmatrix} \tag{6.20}$$

由于 $\boldsymbol{x} = [x_1, x_2]^{\mathrm{T}}$ 和 $\boldsymbol{n} = [n_1, n_2^*]^{\mathrm{T}}$。接收信号可以写成：

$$\boldsymbol{r} = \begin{bmatrix} r_1 \\ r_2^* \end{bmatrix} = \begin{bmatrix} h_1 & h_2 \\ -h_2^* & h_1^* \end{bmatrix} \cdot \begin{bmatrix} x_1 \\ x_2 \end{bmatrix} + \begin{bmatrix} n_1 \\ n_2 \end{bmatrix} \tag{6.21}$$

因此，可得到：

$$r_1 = h_1x_1 + h_2x_2 + n_1 \tag{6.22}$$

$$r_2^* = h_1^*x_2 - h_2^*x_1 + n_2^* \Rightarrow r_2 = h_1x_2^* - h_2x_1^* + n_2 \tag{6.23}$$

和式(6.14)一样。解码过程是：

$$\hat{\boldsymbol{x}} = \boldsymbol{C}^H\boldsymbol{r} = \begin{bmatrix} h_1^* & -h_2 \\ h_2^* & h_1 \end{bmatrix} \begin{bmatrix} r_1 \\ r_2^* \end{bmatrix} = \begin{bmatrix} h_1^*r_1 - h_2r_2^* \\ h_2^*r_1 + h_1r_2^* \end{bmatrix} \tag{6.24}$$

它的形式与式(6.17)和式(6.18)相同。也就是说可以用公式(6.19)—(6.24)这种更简单的形式来表示 Alamouti 空时分组码。

空时分组编码简单的线性操作意味着它不是真的编码，仅能产生分集增益，不能获得真正的编码增益。Tarokh 等人把 Alamouti 方案推广到任意数目的发射和接收天线的系统中。本章节采用四根发射天线发送复数信号 $\mathcal{G}_4$,编码过程为:

$$\mathcal{G}_4 = \begin{bmatrix} x_1 & x_2 & x_3 & x_4 \\ -x_2 & x_1 & -x_4 & x_3 \\ -x_3 & x_4 & x_1 & -x_2 \\ -x_4 & -x_3 & x_2 & x_1 \\ x_1^* & x_2^* & x_3^* & x_4^* \\ -x_2^* & x_1^* & -x_4^* & x_3^* \\ -x_3^* & x_4^* & x_1^* & -x_2^* \\ -x_4^* & -x_3^* & x_2^* & x_1^* \end{bmatrix} \tag{6.25}$$

其中 x_1, x_2, x_3 和 x_4 是在 8 个连续周期内从 4 根发射天线发送的四个复数符号。这种方式可以达到最大分集阶数,但它的速率只有 1/2。

通过文献[5]和[16]中的推导和式(6.19)—(6.24),可以重新得到多根接收天线情况下通过 $\mathcal{G}_4$ 编码的接收信号是:

$$\boldsymbol{r} = \boldsymbol{Cx} + n \tag{6.26}$$

其中 $\boldsymbol{x} = [x_1, x_2, x_3, x_4]^{\mathrm{T}}$ 是传输的符号向量,信道矩阵 $\boldsymbol{C}$ 表示为:

$$\boldsymbol{C} = [\boldsymbol{C}_1, \cdots, \boldsymbol{C}_{n_{\mathrm{R}}}]^{\mathrm{T}} \tag{6.27}$$

$\boldsymbol{C}_j$ $(j = 1, \cdots, n_{\mathrm{R}})$ 是:

$$\boldsymbol{C}_j = \begin{bmatrix} \boldsymbol{h}_{12,j} & \boldsymbol{h}_{34,j} \\ \boldsymbol{h}'_{34,j} & -\boldsymbol{h}'_{12,j} \\ \boldsymbol{h}^*_{12,j} & \boldsymbol{h}^*_{34,j} \\ -\boldsymbol{h}'^*_{34,j} & \boldsymbol{h}'^*_{12,j} \end{bmatrix} \tag{6.28}$$

其中 $\boldsymbol{h}_{mn,j} = \begin{bmatrix} h_{m,j} & h_{n,j} \\ h_{n,j} & -h_{m,j} \end{bmatrix}$, $\boldsymbol{h}'_{mn},j = \begin{bmatrix} h_{m,j} & -h_{n,j} \\ h_{n,j} & h_{m,j} \end{bmatrix}$ 。

接收信号和 *AWGN* 噪声向量是:

$\boldsymbol{r} = [\boldsymbol{r}_1, \cdots, \boldsymbol{r}_{n_{\mathrm{R}}}]^{\mathrm{T}}$,其中 $\boldsymbol{r}_j = [r_{1,j}, \cdots, r_{4,j}, r^*_{5,j}, \cdots, r^*_{8,j}]^{\mathrm{T}}(j = 1, \cdots, n_{\mathrm{R}})$, $\boldsymbol{n} = [n_1, \cdots, n_{n_{\mathrm{R}}}]^{\mathrm{T}}$, $\boldsymbol{n}_j = [n_{1,j}, \cdots, n_{4,j}, n^*_{5,j}, \cdots, n^*_{8,j}]^{\mathrm{T}}$ 。

$h_{i,j}(i = 1,2,3,4; j = 1, \cdots, n_{\mathrm{R}})$ 表示公式(6.2)-(6.4)中定义的 4 根发射天线和 n_{R} 根接收天线之间的信道。$r_{1,j}, \cdots, r_{8,j}(j = 1, \cdots, n_{\mathrm{R}})$ 表示在 8 个周期内从 n_{R} 根接收天线得到的信号。

接收信号通过线性变换:

$$\hat{\boldsymbol{x}} = \boldsymbol{C}^{\mathrm{H}}\boldsymbol{r} = \boldsymbol{C}^{\mathrm{H}}\boldsymbol{Cx} + \boldsymbol{n}' \tag{6.29}$$

把式(6.17)-式(6.20)代入上式,得到 x_1 的估计信号 $\hat{x}_1$:

$$\hat{x}_1 = 2\sum_{j=1}^{n_R}\sum_{i=1}^{n_T}\left|h_{i,j}\right|^2 x_1 + \Xi_1 \tag{6.30}$$

其中 $\boldsymbol{\Xi} = [\Xi_1,\cdots,\Xi_{n_T}]^T$ 且

$$\boldsymbol{\Xi} = \boldsymbol{C}^H \boldsymbol{n} \tag{6.31}$$

对应 $h_{i,j}, i = 1,2,3,4; j = 1,2,\cdots,n_R$;Ξ_1 是零均值复数高斯随机变量,实数方差为 $(4/SNR)\sum_{j=1}^{n_R}\sum_{i=1}^{4}\left|h_{i,j}\right|^2$,信号功率为 $4\left[\sum_{j=1}^{n_R}\sum_{i=1}^{4}\left|h_{i,j}\right|^2\right]^2$。

和 1 根发射天线发射 x_1,$4n_R$ 根天线接收的情况进行比较,x_1 的估计信号 $\hat{x}'_1$ 是:

$$\hat{x}'_1 = \sum_{j=1}^{n_R}\sum_{i=1}^{n_T}\left|h_{i,j}\right|^2 x_1 + \Xi'_1 \tag{6.32}$$

其中 Ξ'_1 是零均值实数方差是 $(1/\mathrm{SNR})\sum_{j=1}^{n_R}\sum_{i=1}^{4}\left|h_{i,j}\right|^2$ 的随机复高斯变量。这和式(6.22)中的形式相同,只是信号和噪声的功率都降低了 1/4。

因此,使用 $\mathcal{G}_4$ 编码可以提供与 $4n_R$ 级最大比合并(MRC)完全相同的性能。

图 6-4 显示了使用 STBC 的 MIMO 系统的传输 QPSK 调制信号的仿真结果。仿真是逐帧进行的。

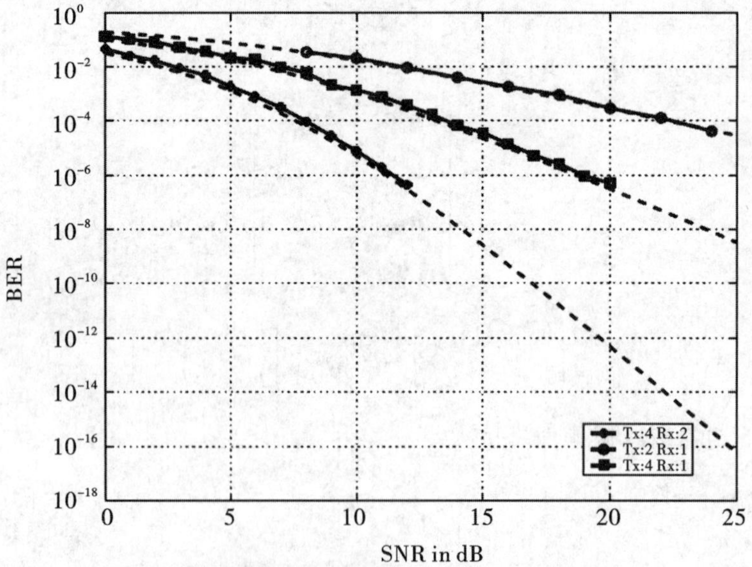

图 6-4 通过 STBC 编码的不同数量发射和接收天线 MIMO 系统的 BER 性能

实线表示阶数为 2,4,8 的 BER 仿真性能,虚线表示理论性能

信道是式(6.2)和式(6.3)所示的独立瑞利信道,噪声是 AWGN。信道假定是准静态的,即在多个空时编码块的持续时间内保持静止。在 STBC 解码算法中,接收端需要进行

信道估计,因为接收端需要了解完全的信道状态信息。结果表明,随着发射和接收天线数目的增加,所获得的分集阶数也随之增加。2 根发射天线和 1 根接收天线(Tx:2,Rx:1)的 MIMO 系统,使用 Alamouti 方案,可以提供 2(bits/s)/Hz 的传输速率,对于 4 根发射天线和 1 根接收天线(Tx:4,Rx:1)的系统,其传输速率为 1 (bit/s)/Hz。对于 4 根发射天线和 2 根接收天线(Tx:4,Rx:2)的 MIMO 系统,其传输速率同样为 1 (bit/s)/Hz。仿真也对公式(6.5)-式(6.12)的分析结果进行了比较。这些结果很好地应证了 STBC 可以实现 $n_R n_T$ 的满分集。

6.5 空间复用

在空间复用(SM)系统中,要发送的数据流首先被分解,然后从每根天线同时发送。在接收机处,每根天线观察来自所有发射天线发射的信号的叠加,然后利用适当的信号处理技术将它们分离并进行复用以恢复原始数据流。图 6-5 显示了一个 SM 系统。使用多天线检测器分离接收到的信号,该多天线检测器执行信号处理的过程类似于多用户检测器中分离用户的过程。前向纠错(FEC)编码几乎总是与 SM 方案相结合来提高性能。通常子流是分开编码的,但也可以使用联合编码。

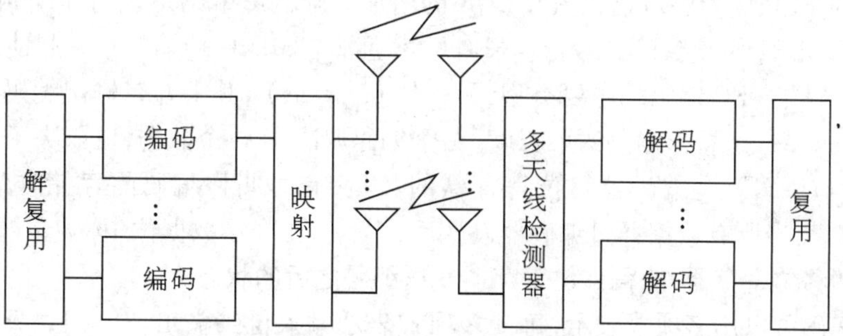

图 6-5 空间复用系统示意图

SM 方案中最重要的例子是 BLAST(贝尔实验室分层空时码结构)技术,D-BLAST 和 V-BLAST(分别是对角和垂直)。前者涉及天线使用编码和变量映射,而后者(原则上)则不。

ML 检测算法用于最小化 BER 并提供最优性能的 SM 检测算法。然而,这种技术的主要缺点是计算复杂性。因为它必须进行 M^{n_T} 次矢量搜索,其中 M 是一系列中符号的数目(例如对于 QPSK, $M=4$)。因此,为了降低检测器的复杂度,可以考虑使用次最优技术,例如迫零(ZF) 和 MMSE 的线性处理技术。

ZF 方法用于估计发送的符号,如下所示:

$$\hat{\boldsymbol{x}} = \boldsymbol{H}^{\mathrm{H}} \left(\boldsymbol{H} \boldsymbol{H}^{\mathrm{H}}\right)^{-1} \boldsymbol{r} \tag{6.33}$$

接收机在高 SNR 情况下工作性能良好,而在低 SNR 情况下会有严重的噪声增强。

另一种可选择的线性接收机是 MMSE,它最小化价值成本函数 $E\{|x-\hat{x}|\}$ 来估计一

个随机变量,其中 $E\{\cdot\}$ 代表期望值。在这种情况下, x 的估计值由下式给出:

$$\hat{\boldsymbol{x}} = (\boldsymbol{H}^{\mathrm{H}}\boldsymbol{H} + \sigma^2\boldsymbol{I})^{-1}\boldsymbol{H}^{\mathrm{H}}\boldsymbol{r} \tag{6.34}$$

其中 $\boldsymbol{I}$ 是 $n_{\mathrm{T}} \times n_{\mathrm{T}}$ 的单位矩阵。在高 SNR 区间,MMSE 接收机接近 ZF 接收机。MMSE 接收机以降低分离信号质量为代价,抗噪声干扰能力较强[2]。

同时也有例如排序连续干扰消除(OSIC)非线性技术,OSIC 是文献[18]、[19]中提出的初始解码算法。它可以从具有最高 SNR 的接收天线中检测接收信号,估计发送符号,然后把这个符号从接收的信号中去除从而检测剩余的发送信号。这个过程会一直重复直到估计出所有的发送符号。这个算法通过处理可以获得软输出。

如果 n_{T} 等同于多用户检测中的用户数 K , n_{R} 等同于扩频因子,空间复用 MIMO 系统检测类似于多用户检测。矩阵 $\boldsymbol{H}^{\mathrm{H}}\boldsymbol{H}$ 等同于在多用户检测的扩频序列相关矩阵 $\boldsymbol{R}$。

6.6 卷积码

FEC 编码是通过在发送数据位中插入检查位或奇偶校验位来改善通信系统 BER 性能的技术。奇偶校验位可以纠正由信道引起的接收信号中的错误,并有助于重建传输信息。很明显,BER 随着解调器输入端 SNR 的增加而降低。编码系统与非编码系统相比,取得相同的 BER,所需要 SNR 较低。FEC 技术通常与 MIMO 传输技术一起使用。

卷积编码是 FEC 编码的主要类型,形式是 (l_s,k_s,v_i) ,其中 l_s 是输出编码长度, k_s 是输入数据长度, v_i 是卷积编码约束长度。l_s 个编码取决于 k_s 个数据,也取决于前 (v_i-1) 个数据块。(v_i-1) 是编码器移位寄存器的数量,有时叫作编码的存储顺序。码率是 $k_s = l_s$ 。注意,当所有数据通过编码器发送后, (v_i-1) 个零数据位(称为尾部位)被发送到编码器的移位寄存器,使得移位寄存器发送编码之后的状态为零。

它的编码可以用多项式表示,每个多项式表示输入位到输出位。(2,1,3)卷积编码器的结构如图 6-6 所示,对应的多项式是:

$$g^{(1)}(D) = D^2 + D + 1 \tag{6.35}$$

$$g^{(2)}(D) = D^2 + 1 \tag{6.36}$$

其中 D 代表编码器中的移位寄存器。多项式中 D 的最高次幂表示编码器移位寄存器的数目,在式(6.35)和式(6.36)中, $D=2$。多项式中 D^i 项意味着第 i 个移位寄存器的状态 S_{i-1} 的对应编码和模 2 加法器相连。

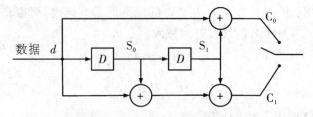

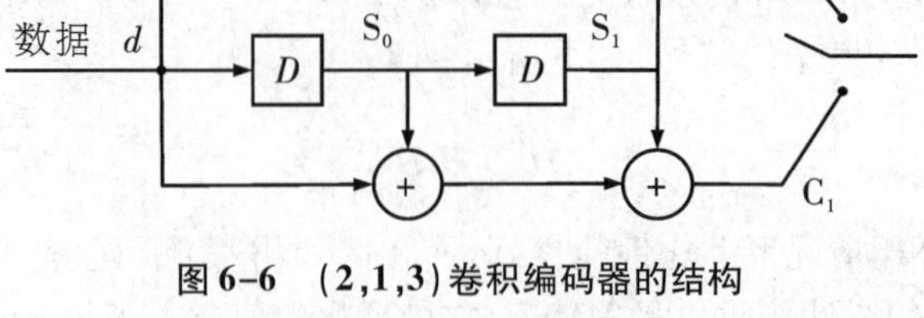

图 6-6 (2,1,3)卷积编码器的结构

6.7 MIMO与多用户检测的硬件实现

从Foschini、Gans、Teletar和Paulraj的早期研究开始，许多关于MIMO的信息理论、算法、代码等领域的论文被发表了。多数研究集中在给定SNR的情况下获取更好的BER的算法和协议上。很少研究关注这些算法的实时实现。只有考虑算法的硬件复杂性，才可以方便MIMO检测器与系统的其余部分相集成。到目前为止，MIMO算法通常用数字信号处理器(DSP)实现，但难以实现高数据速率性能。现场可编程门阵列(FPGA)器件因其可重配置性和支持并行处理而广泛用于信号处理、通信和网络应用。

与DSP处理器相比，FPGA至少具有三个优点：具有内在并行性的FPGA可用于向量处理；它减少了指令开销；当FPGA资源足够时，处理容量可以被扩展。它的缺点是开发周期通常比DSP实现要长。但是一旦完成了一个高效的体系结构并且开发了并行实现，它可以显著提高处理速度。

另外，FPGA相对于专用集成电路(ASIC)实现有几个优点：FPGA器件可重新配置，即使在运行时也能适应系统配置的变化；与ASIC相比，延迟明显缩短；它是具有成本效益的解决方案。而且，ASIC实现通常应用于固定数量的天线和特定的信号系统。ASIC实现的局限性是当天线数量或信号系统发生变化时缺少灵活性。

MIMO检测器通常在DSP上实现，如贝尔实验室的分层空时(BLAST)系统。但由于不支持并行计算，DSP实现的速度往往是有限的，特别是随着天线数量的增加。最近在文献[28]—[31]中报道了许多SD或接近ML检测器的ASIC和FPGA实现。使用球形解码算法的深度优先检测器的VLSI设计和使用M-算法的宽度优先检测器都引起了许多关注。对于16-QAM的4×4 MIMO传输，硬输出深度优先检测器实现比其宽度优先检测器有更高的吞吐量。一般来说，深度优先硬输出检测的平均计算复杂度低于对应的宽度优先检测器，因为它具有自适应收紧搜索半径约束的能力。对于软输出检测，自适应地改变深度优先搜索和广度优先搜索的搜索半径是重要的，而广度优先搜索的优点是它可以自然地生成一个有序候选者列表，用于后验概率(APP)计算。

表6-1比较了4×4 16-QAM MIMO检测算法的相关硬件实现，到目前为止，MIMO通信的大多数硬件实现都是处理小规模系统。深度优先的SD1和SD2是基于ASIC的实现。SD2实现占SD1芯片的一半面积，吞吐量是SD1的两倍。深度优先树搜索是以顺序和非流水线方式实现的，而K最佳算法是基于并行和流水线，减少了芯片面积。此外，K最佳算法保证了恒定的吞吐量，但是以性能损失为代价。深度优先方法的平均吞吐量可以与K最佳算法的平均吞吐量相匹配，但在最坏情况下，其吞吐量可能会严重下降。FPGA SD3与SD1有类似的性能，但其复杂性显著降低。还需要提及的是，表6-1中SD的实时实现不包括会消耗更多硬件资源的QR分解或Cholesky分解等信道矩阵预处理。

表 6-1　4×4 16 QAM MIMO 检测的实现比较

技术	BER	硬件平台	时钟频率	在 SNR = 20 dB 的吞吐量(Mbps)	FPGA 逻辑片数目或者 ASIC 门数目
K-best 1[35,28]	close ML	ASIC	100 MHz	10	52000
K-best 2[35,29]	close ML	ASIC	100 MHz	52	91000
Depth-first SD1 (ℓ^2-norm)[31,30]	ML	ASIC	51 MHz	73	117000
Depth-first SD2 (ℓ^1-norm)[31,30]	close ML	ASIC	71 MHz	169	50000
Depth-first SD3[31]	ML	FPGA	50 MHz	114.5	21467

多用户检测器的实现更加困难,因为它们主要用于大规模系统。这里有一些参考文献,例如在[35]中介绍了一种自适应 MMSE 算法的 FPGA 实现,在[36]中介绍了基于自适应滤波器级联的 FPGA 多用户检测器的异步 WCDMA 系统。

第7章 基于盒型约束 DCD 的多用户检测

本章介绍基于盒型约束的二分坐标下降(DCD)多用户检测方法。二分坐标下降算法不需要乘法和除法操作,因此非常适合硬件实施。DCD 算法的 FPGA 架构设计有顺序执行和并行执行两种。把 DCD 定点型算法写入 FPGA 板子,结果显示即使在大规模用户系统中,DCD 算法优于 MMSE 算法,而且硬件资源占用量非常小。在保证检测精度的同时,DCD 的并行设计比顺序执行设计的数据传输吞吐量高。

7.1 概述

在 CDMA 系统中,多用户检测能够提供高检测性能。使用多用户检测技术可以显著提高无线通信系统的频谱效率。因此该项技术已经发展成为多址访问通信的一个重要研究领域。当多用户信号能量不同或者用户多的情况下,传统的检测器(匹配滤波器)表现不佳。ML 检测器复杂度随用户数增多而呈指数增长,使其在实时操作中不可行。球形解码算法可以简化 ML 多用户检测。然而,如 Cholesky 或 QR 等的矩阵分解的硬件实现是非常困难的。因此,球形解码算法仅适用于小型系统。

盒型约束二分坐标下降(DCD)算法不需要乘法和除法运算来求解一般方程,使其更适合于实时实现。本章我们主要学习盒型约束 DCD 算法。盒型约束 DCD 多用户检测器非常适用于大规模检测系统。

FPGA 可以实现盒型约束 DCD 算法。本章将介绍两种 FPGA 架构设计。盒型约束 DCD 算法的串行架构比球形解码器的 FPGA 实现的复杂度小得多,因为它没有直接的乘法和除法,只需要加法和位移操作。通常情况并行架构设计比串行设计的数据吞吐量高。因此,除了串行架构设计,我们还将介绍盒型约束 DCD 检测器的并行 FPGA 设计。

7.2 多用户检测问题

我们考虑 AWGN 信道中使用 BPSK 调制的 K 个用户同步 CDMA 系统。接收器通过匹配滤波器输出:

$$\boldsymbol{\theta} = \boldsymbol{R}\boldsymbol{x} + \boldsymbol{n} \tag{7.1}$$

其中向量 $\boldsymbol{x} \in \{-1, +1\}^K$ 包含 K 个用户发送的比特信息,$\boldsymbol{R}$ 是 $K \times K$ 实数矩阵,$\boldsymbol{x}$ 和 $\boldsymbol{\theta}$ 是 $K\times1$ 实数向量,$\boldsymbol{n}$ 是具有协方差矩阵 $\sigma^2\boldsymbol{R}$ 的实数零均值高斯随机向量。最优 ML

多用户检测器使用整数约束最小化下面的二次价值成本函数来估计向量 $\boldsymbol{x}$：

$$\hat{\boldsymbol{x}} = \underset{x \in \{-1,+1\}^K}{\operatorname{argmin}} \left\{ \frac{1}{2} \boldsymbol{x}^{\mathrm{T}} \boldsymbol{R} \boldsymbol{x} - \boldsymbol{\theta}^{\mathrm{T}} \boldsymbol{x} \right\} \tag{7.2}$$

尽管 ML 检测器可以提供最佳的检测性能，但由于它的高运算量所以并不实用。球形解码器提供和 ML 检测器相同的性能，并显著降低了平均复杂度。然而，在低信噪比情况下，球形解码器最坏情况的复杂度与用户数指数成正比，这就妨碍了在大规模用户系统中使用球形解码器。

7.3 盒型约束 DCD 算法

盒型约束 DCD 多用户检测器在最小化二次价值成本函数（7.2）中使用了盒型约束 $\boldsymbol{x} \in [-1, +1]^K$。表 7-1 给出了盒型约束 DCD 算法。盒型约束 DCD 算法被设计为免直接乘法和除法来求解向量 $\boldsymbol{x}$ 的一种简单方案。解向量 $\boldsymbol{x}$ 的最终准确度取决于比特数（M_b）、迭代次数以及系统矩阵数的条件数等因素。算法中的第一组迭代使用步长参数 d 来确定 $\boldsymbol{x}$ 的最重要的有效位（$m = 1$）。随后的迭代会根据合适精度来确定较低有效位（最大到 M_b）。$\boldsymbol{r} = \boldsymbol{\theta} - \boldsymbol{R}\bar{\boldsymbol{x}}$ 是初始残差矢量，其中 $\bar{\boldsymbol{x}}$ 是 $\boldsymbol{x}$ 的初始化数值。$\bar{\boldsymbol{x}}$ 被设置为零，$\boldsymbol{r}$ 等于 $\boldsymbol{\theta}$。在步骤 1 中，步长 d 通过 2 的指数减少，因此，它不需要乘法或除法，所有乘法和除法都可以用简单的位移代替。如果在一次迭代中解向量被更新，则这样的迭代被标记为“成功”。对于每次步长更新，算法重复成功的迭代，直到残差向量 $\boldsymbol{r}$ 里所有元素变得非常小，使得所有 j 元素不满足步骤 4 的条件或者 x 溢出步骤 6 的范围 $[-H, +H]$，其中对于 BPSK 调制，$H = 1$。该算法的运算量主要取决于成功的更新迭代次数 N_u 和比特数 M_b。预先设定的成功迭代次数 N_u 可以作为算法停止的条件（在步骤 10）。如果没有这样预设，或者预设数值非常高，则解向量的精度是 2^{-M_b+1}。

表 7-1　盒型约束 DCD 算法

初始化：$\boldsymbol{x} = \bar{\boldsymbol{x}}$，$\boldsymbol{r} = \boldsymbol{\theta} - \boldsymbol{R}\bar{\boldsymbol{x}}$，$H = 1$，$p = 0$.

for $m = 1 : M_b$

1. $d = 2^{-m+1}$
2. **Flag** = 0　　　　…pass
3. **for** $j = 1 : K$　　　　…iteration
4. 　**if** $|r(j)| > (d/2)R(j,j)$ **then**
5. 　　$x = x(j) + \mathrm{sign}(r(j)) \cdot d$
6. 　　**If** $|x| < H$ then
7. 　　$x(j) = x$
8. 　　$\boldsymbol{r} = \boldsymbol{r} - \mathrm{sign}(r(j)) \cdot d \cdot \boldsymbol{R}(:,j)$
9. 　　**Flag** = 1，$p = p + 1$
10. 　　**if** $p > N_u$ **end algorithm**
11. **end** j**-loop**
12. **if Flag** = 1，go to 2

end m-loop

一次成功迭代需要一个用于比较的加法(在步骤 4),以及更新残差向量 $\boldsymbol{r}$ 和元素 $x(j)$ 需要的 $(K+1)$ 个加法。如果是一次不成功的迭代,只需要用于比较的一个加法。最坏情况下的复杂度是计算一种不太可能出现的情形,即当只有最后第 m 位有 N_u 次成功迭代时才发生。这意味着前 (M_b-1) 位的计算不包含任何成功迭代,因此需要 $(M_b-1)K$ 个加法。计算最后一位(对应于 $m=M_b$)的最坏情况出现在 K 次迭代 $(j=1,\cdots,K)$ 中只发生一次成功迭代。这就需要用 K 个加法来比较,$(K+1)$ 个加法来更新残差向量 $\boldsymbol{r}$ (在步骤 8)和元素 $x(j)$ (在步骤 5)。总的来说,N_u 次成功迭代需要 $N_u(2K+1)$ 个加法。

因此,盒型约束 DCD 算法的复杂度的上限是 $K(2N_u+M_b-1)+N_u$ 次加法。然而,在通常情况下,每位计算都应该会有几次成功迭代,平均复杂度接近于 $2KN_u$.

图 7-1 为 DCD 处理器的框图。

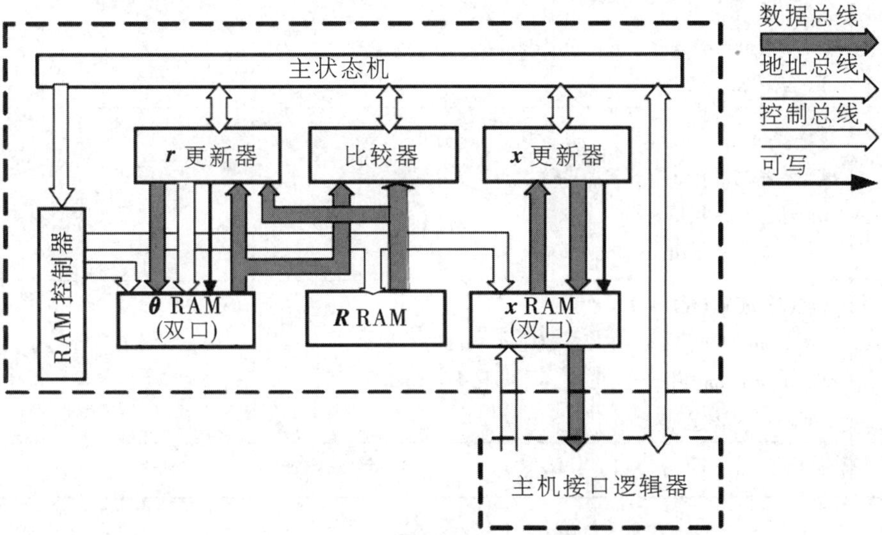

图 7-1　DCD 处理器的框图

7.4　盒型约束 DCD 算法的定点型串行架构

可以用 VHDL 描述定点型盒型约束 DCD 算法并把它写入 FPGA。开发板带有 FPGA 芯片 XC2VP30(FFT896 封装,速度等级 7)的 Xilinx Virtex-II Pro 开发系统。定点型 DCD 算法在 100 MHz 时钟频率下的 Xilinx ISE 8.1i 运行、综合之后下载到该 FPGA 芯片。定点型设计使用 16 位 Q15 数字格式来表示矩阵 $\boldsymbol{R}$ 的元素。为了避免溢出错误,可以使用 32 位定点型整数来表示向量 $\boldsymbol{\theta}$ 和 $\boldsymbol{x}$ 中的元素。这些元素被限制在 $[-2^{16},2^{16})$·范围内。我们把存储在 $\boldsymbol{\theta}$ RAM 中的向量作为残差向量 $\boldsymbol{r}$。

表 7-2 描述了定点型盒型约束 DCD 算法的实数串行执行步骤。图 7.2 显示了定点型盒型约束 DCD 算法在六种状态之间的转换。图 7-1 显示了 DCD 处理器的框图。向量 $\boldsymbol{x}$ 和 $\boldsymbol{r}$ 分别存储在 $\boldsymbol{x}$RAM 和 $\boldsymbol{\theta}$ RAM 中。在状态 0,初始化向量 $\boldsymbol{x}$,$\boldsymbol{r}$,位计数器 m ,成功的迭代计数器 p ,预设的缩放计数器 Δm ,元素索引 j 和 $Flag$。在状态 1 中,如果 $m\neq 0$,则步长 $d=2^m$,m 减 1,Δm 加 1。如果获得了最低有效位($m=0$),则算法停止。

表 7-2 定点实值盒型约束 DCD 算法

状态	操作	周期数
0	初始化: $x=0$, $\boldsymbol{r}=\boldsymbol{\theta}$, $m=M_b$, $p=0$ $\Delta m=0$, $j=1$, Flag$=0$	
1	if $m=0$, algorithm stops else, $m=m-1$, $d=2^m$, $\Delta m=\Delta m+1$	1
2	$c=R(j,j)/2-\mid r(j)\mid\times 2^{\Delta m}$ $x_t=x(j)+\text{sign}(r(j))\cdot d$ $\mid x_t\mid\leqslant H$, then $\alpha=0$; else $\alpha=1$	1
3	if $c<0$ and $\alpha=0$, then goto state 4 else, goto state 5	1
4	$x(j)=x_t$ $\boldsymbol{r}=\boldsymbol{r}\times 2^{\Delta m}-\text{sign}(r(j))\cdot\boldsymbol{R}(:,j)$ $\Delta m=0$, $p=p+1$, Flag$=1$ if $p=N_u$, algorithm stops	K
5	$j=(j)\bmod(K)+1$ if $j=1$, and Flag$=1$ then Flag$=0$ goto state 2 else if $j=1$ and Flag$=0$, then goto state 1 else, goto state 2	1
Total	$\leqslant 4KN_u+3K(M_b-1)+M_b$	

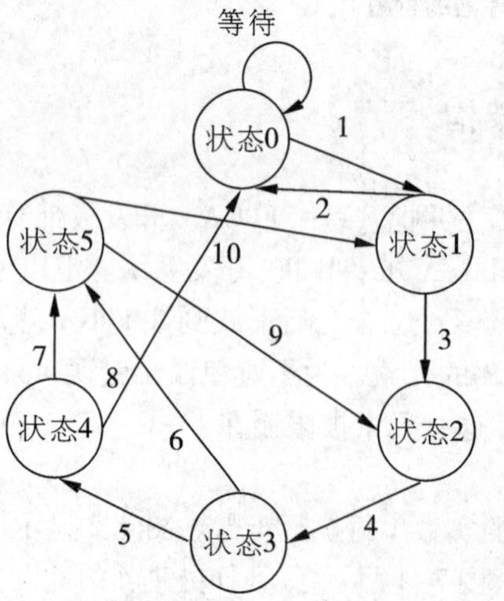

1:开始计算
2:m 次循环完成
3:作比较
4:查询比较结果
5:成功迭代,去更新
6:不成功迭代,作下一个元素比较
7:更新计数器 j
8:到达最大迭代数，退出计算
9:Flag=0,回到下一次比较
10:第一个元素是不成功迭代,更新步长和 m

状态0:初始化状态
状态1:m 循环控制和步长更新
状态2:比较
状态3:判断是“成功”或者“不成功”
状态4:更新 $x(j)$ 和向量 $\boldsymbol{r}$
状态5:j 循环控制

图 7-2 DCD 主状态机

在状态2中，在每次比较之前需要两个周期，因为从RAM中读取数据有一个时钟延迟。RAM控制器分指定保存在$\boldsymbol{\theta}$ RAM，$\boldsymbol{R}$ RAM和$\boldsymbol{x}$ RAM中的$r(j)$，$R(j,j)$和$x(j)$的地址。同时，$r(j)$用$2^{\Delta m}$比特移位进行缩放。另外，$\boldsymbol{x}$更新逻辑器从$\boldsymbol{x}$ RAM读取$x(j)$，预更新$x(j)$并将更新的元素保存在寄存器x_t中。在进入状态3之前，它还检查x_t是否在$[-H,H]$ $(\alpha=0)$或$(\alpha=1)$的范围内。

在状态3中，主状态机检查来自状态2的比较结果，以判断迭代是否"成功"。如果迭代"成功"，则算法进入状态4来更新$\boldsymbol{x}$和$\boldsymbol{r}$；否则，算法进入状态5。

在状态4中，$\boldsymbol{r}$更新器更新向量$\boldsymbol{r}$的所有元素。RAM控制器指示$\boldsymbol{R}(:,j)$列和向量$\boldsymbol{r}$里面所有元素的地址。$\boldsymbol{r}$和$\boldsymbol{R}(:,j)$里的相应元素被加载到$\boldsymbol{r}$更新器中。$x(j)$通过直接拷贝寄存器x_t的值进行更新。同时，Δm清零，迭代计数器p加1，并且二进制变量Flag被设置为1，这表示当前迭代是"成功"的。接着需要检查"成功"迭代计数器p是否达到了最大数量N_u。如果p小于N_u，主状态机进入状态5，否则算法停止。

状态5首先更新索引j。然后，判断是否更新Flag，并根据j和Flag的数值决定进入哪一个状态。

表7－2列出了每个状态所需的时钟周期数。通过使用流水线技术，状态4需要K个周期来更新向量$\boldsymbol{r}$中的所有元素和$\boldsymbol{x}$中的一个元素。其他状态只需要一个执行周期。因为从RAM中读取数据时存在一个时钟周期延迟，状态2需要2个时钟周期。所需的总周期主要取决于这些成功更新迭代次数和位数。表7－2中周期数的上限可以看作是定点型盒型约束DCD算法的最差情况对应的复杂度。如果考虑最差的情况，即只有最后的第m位有N_u次成功迭代。这意味着前(M_b-1)位的计算不包含任何成功迭代，因此需要$(M_b-1)3K$个周期。计算最后一位($m=1$)的最差情况发生在K次迭代$(j=1,\cdots,K)$中只发生一次成功的迭代。这需要$3K$个周期作比较和K个周期来更新残差向量$\boldsymbol{r}$和$x(j)$元素。N_u次成功迭代总共需要$N_u(4K)$个周期。除了上述所需的周期外，整个过程中步长更新也需要M_b个周期。

因此，定点型盒型约束DCD算法的最大时钟周期个数为$4KN_u+3K(M_b-1)+M_b$。

图7－1给出了盒型约束DCD算法的FPGA架构。这个架构有五个子模块：主状态机，RAM控制器，比较器，$\boldsymbol{r}$更新器和$\boldsymbol{x}$更新器。矩阵$\boldsymbol{R}$、向量$\boldsymbol{r}$和向量$\boldsymbol{x}$分别保存在$\boldsymbol{R}$ RAM，$\boldsymbol{\theta}$ RAM和$\boldsymbol{x}$ RAM中。

主状态机：主状态机需要跟踪当前的迭代，选择和更新步长，并决定执行哪个状态。

RAM控制器：RAM控制器是选取$\boldsymbol{\theta}$ RAM和$\boldsymbol{R}$ RAM的地址端口，以及$\boldsymbol{x}$ RAM的两个地址端口。在一次迭代的"比较"期间，RAM控制器提供$R(j,j)$，$r(j)$和$x(j)$的地址。更新操作是有条件的；如果迭代是"成功的"，那么RAM控制器将顺序递增向量$\boldsymbol{r}$和矩阵$\boldsymbol{R}$中的地址，更新$\boldsymbol{r}$向量中的所有K个元素。同时，RAM控制器指向$\boldsymbol{x}$ RAM中$x(j)$的地址来更新$x(j)$。然后RAM控制器指向下一个比较地址$r(j+1)$和$R(j+1,j+1)$。但是，如果迭代"不成功"，RAM控制器立即指向地址$r(j+1)$和$R(j+1,j+1)$（不更新）用于下一次比较。

比较器：比较器架构如图7－3所示。它执行表7－1中的步骤4中的$r(j)$和$(d/2)R(j,j)$的比较。比较器还缩放$r(j)$来抵消表7－2中状态2的d，并将结果c的符号位传递给主状态机。

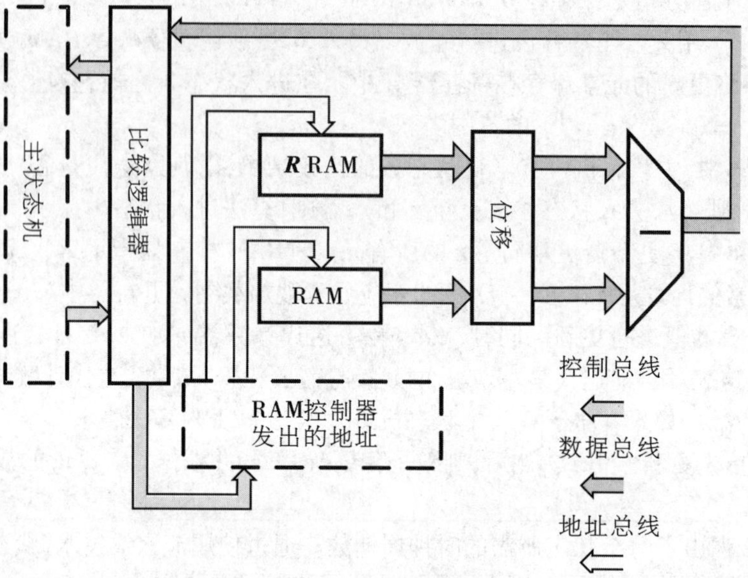

图 7－3　比较器

r 更新器:图 7－4 给出了 **r** 更新器的架构。向量 **r** 和 **x** 初始化值被同时分别存储在 **θ** RAM 和 **x** RAM 中。**θ** RAM 储存了调整后的数据,而 **x** RAM 被清零。每个元素都从 **θ** RAM 中被读取,按照 $\boldsymbol{r}=\boldsymbol{r}\times 2^{\Delta m}-\mathrm{sign}(\mathrm{r}(\mathrm{j}))\cdot \boldsymbol{R}(:,j)$ 更新,并写回到 **θ** RAM 中。

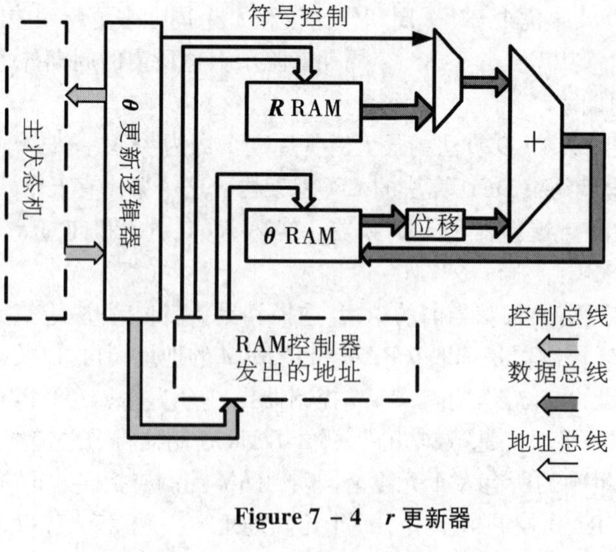

Figure 7－4　r 更新器

x 更新器:图 7－5 显示了 **x** 更新器。**x** 更新器从 **x** RAM 中读取元素 $x(j)$ 。$x(j)$ 根据 $r(j)$ 的符号加上或减去步长 d 来更新 $x(j)$。更新的 $x(j)$ 被写回到 **x** RAM 中。

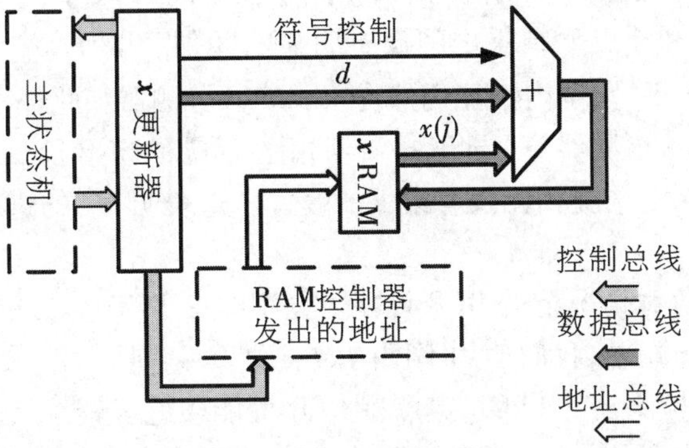

图7-5 x 更新器

7.4.1 盒型约束DCD的检测性能

我们通过10^5次模拟试验评估盒型约束DCD检测器的BER性能。假设仿真环境是完美功率控制的AWGN信道,用户采用随机生成的扩频序列。考虑用户数$K=50$,扩频因子$SF=53$的高负载情况,以及$K=20$和$SF=31$的低负载情况。我们还研究了在这些情况下该检测器的定点型和浮点型的表现。浮点型执行结果由MATLAB获得,定点型执行结果由FPGA芯片获得。

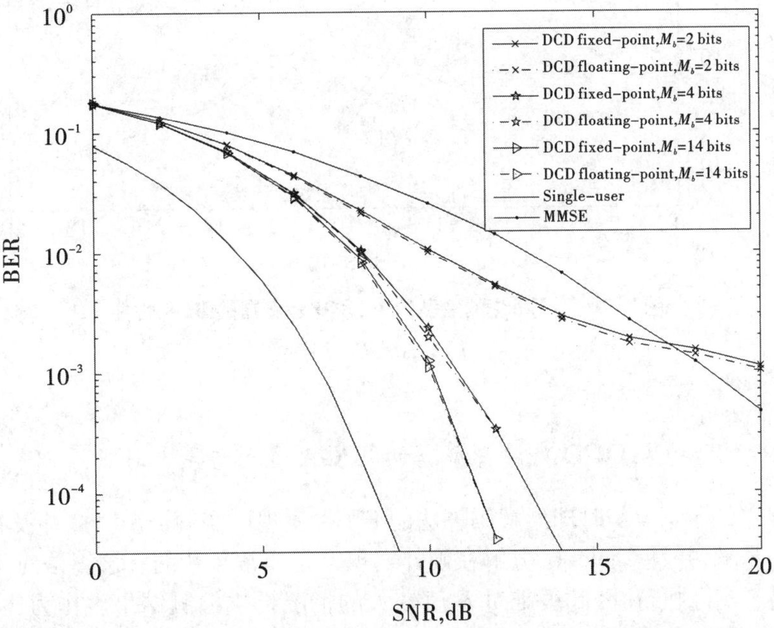

图7-6 高负载多用户情况下盒约束DCD多用户检测器不同M_b的BER性能

$K=50, SF=53$

图 7-6 显示了检测器在高负载情况下不同 M_b 在不同 SNR 下的 BER 性能。可以看出，定点型 FPGA 结果和浮点型结果之间没有明显的区别。即使 $M_b = 2$，盒型约束 DCD 检测器在 0 dB 到 17.5 dB 范围内性能仍然优于 MMSE 检测器。此外，当 BER = 10^{-3} 时，$M_b = 4$ 的盒型约束 DCD 检测器与 MMSE 检测器相比，所需的 SNR 降低了 7.5 dB。$M_b = 14$ 时，BER 性能与 $M_b = 4$ 的情况相比没有太大的改善。这表明 M_b 的进一步增加将不会显著改进 BER 性能。

图 7-7 显示了低负载情况下的 BER 性能。盒型约束 DCD 算法中的参数 M_b 可以选择任何整数。在这个仿真中，我们使用不同的 M_b（即 2、4、14）来演示它的检测性能。结果表明盒型约束 DCD 检测器的 FPGA 定点型 BER 性能接近于其浮点型结果。即使 $M_b =$ 2 的盒型约束 DCD 检测器在 0 dB 至 20 dB 的 SNR 范围内优于 MMSE 检测器。此外，图 7-7 还表明，$M_b = 4$ 的盒型约束 DCD 检测器的检测性能几乎与 $M_b = 14$ 的检测性能相似。因此，对于盒型约束 DCD 检测器而言，仅 $M_b = 4$ 就足够的。

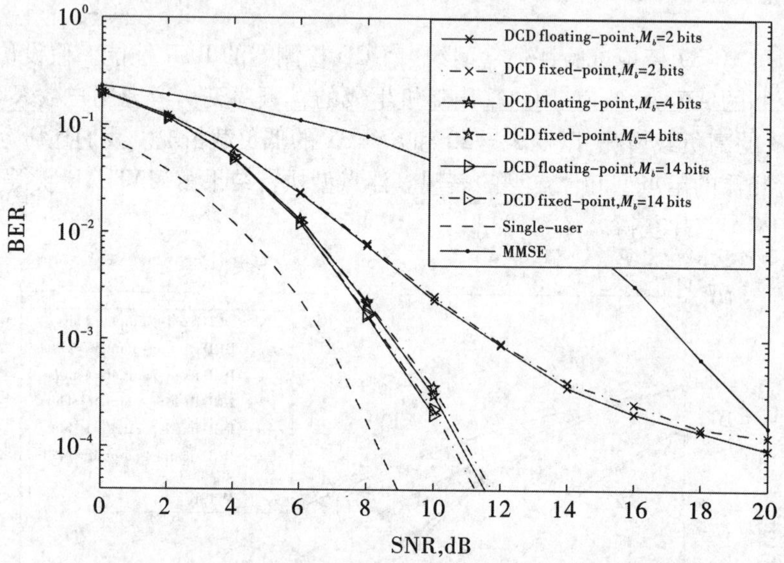

图 7-7　不同 M_b 的盒型约束 DCD 检测器的 BER 性能

$K = 20, SF = 31$

7.4.2　串行盒型约束 DCD 检测器所需的 FPGA 资源与最大时钟周期

表 7-3 总结了 $K = 4$ 时球形解码算法的 FPGA 实现所需的资源。这并没有考虑伪逆所需要的运算量，并且 Cholesky 分解在硬件实现上是非常复杂的。该算法对于大规模用户来说不易于实际执行，因此很难对 K 值较大的情况提供检测结果。作为比较，表 7-4 总结了 $K = 50$ 和 $M_b = 4$ 时盒型约束 DCD 算法所需的 FPGA 资源。该算法的用户数比球形解码算法的用户数多很多，所以它们并不是等同的比较；然而，由于球形检测器可以提供最佳的检测性能，它仍然是值得比较的。在盒型约束 DCD 算法实现中使用的乘法器数为零，因为它的乘法操作是通过简单的位移操作实现的。在球形解码算法实现中使用

的逻辑资源数量大约是盒型约束 DCD 算法中使用逻辑资源的 3 倍。此外，盒型约束 DCD 检测器执行中使用的 RAM 数量比球形解码器少了约 20 倍。我们也实现了盒型约束 DCD 检测器在 $K=4$ 时的硬件实施（这里未显示结果）。我们发现当改变系统大小（即 K 从 4 到 50）时，盒型约束 DCD 算法使用的逻辑资源不会有很大的变化，但是当系统规模增加时，RAM 数将会增加。

表 7-3　球形解码算法所需的 FPGA 资源（$K=4$）

FPGA 资源	可用资源	使用资源
逻辑片	33088	12721
乘法器	328	160
块 RAM	328	82

表 7-4　盒型约束 DCD 算法所需的 FPGA 资源（$K=50$，$M_b=4$）

FPGA 资源	可用资源	使用资源
逻辑片	13696	387
乘法器	136	0
块 RAM	136	4

表 7-5 显示了不同 K 和不同 M_b 情况下，盒型约束 DCD 算法所需要的最大时钟周期数。对于相同的 K 值，它所需的最大时钟周期数随着 M_b 的增加而增加。对于相同的 M_b 值，所需的最大时钟周期数随着 K 的增加而增加。总之，更新时间随着 M_b 或 K 的增加而增加。

表 7-5　最坏情况下不同 M_b 的盒型约束 DCD 算法所需的时钟周期数

	$K=4$	$K=20$	$K=50$	$K=110$
$M_b=2$	$16N_u+14$	$80N_u+62$	$200N_u+152$	$440N_u+332$
$M_b=4$	$16N_u+40$	$200N_u+454$	$200N_u+454$	$440N_u+994$
$M_b=14$	$16N_u+170$	$200N_u+1964$	$200N_u+1964$	$440N_u+4304$

7.5　盒型约束 DCD 算法的定点型并行架构

已经证明，即使针对大规模用户，盒型约束 DCD 算法的串行执行架构需要很少的硬件资源。但是串行执行操作，使得这种算法架构可能无法提供令人满意的数据吞吐量。目前存储在 $\boldsymbol{\theta}$ *RAM* 中的向量 $\boldsymbol{r}$ 元素被顺序更新。而在本节中，我们将改变向量 $\boldsymbol{r}$ 元素的存储方式。把这些元素存储在寄存器中，这就允许在每一次“成功”迭代中 $\boldsymbol{r}$ 里面所有 K

个元素会在一个时钟周期内被更新。另外,可通过两种方法同时获取列 $\boldsymbol{R}(:,j)$ 中的所有元素。一种方法是把列 $\boldsymbol{R}(:,j)$ 的元素存储在寄存器中时,这种方法叫做 $\boldsymbol{R}$-in-Register。另一种是,把列 $\boldsymbol{R}(:,j)$ 的元素存储在 RAM 中,这种方法叫作 $\boldsymbol{R}$-in-RAM。

表 7-6 给出了并行架构的盒型约束 DCD 算法。在状态 0 中,控制信号被初始化。向量 $\boldsymbol{r}$ 的元素被存储在寄存器中,矩阵 $\boldsymbol{R}$ 的元素被存储在寄存器或 RAM 中。在状态 1 中,如果 $m \neq 0$,在一个周期里 $\boldsymbol{r}$ 里所有元素向左移一位。同时,更新步长 d,m 减 1。在状态 2 中,主状态机将 $\boldsymbol{R}(:,j)$ 里的元素和向量 $\boldsymbol{r}$ 发送给 $\boldsymbol{r}$ 更新器。大约需要两个时钟周期访问这些元素和计算 $\boldsymbol{r}_t$。在状态 3 中,主状态机比较 $\boldsymbol{R}(j,j)$(右移)和 $r(j)$。另外,$\boldsymbol{x}$ 更新逻辑器从 $\boldsymbol{x}$ RAM 读取 $x(j)$,预更新 $x(j)$ 并将更新的元素保存在寄存器 $\boldsymbol{x}_t$ 中。在进入状态 4 之前,它还要检查 $\boldsymbol{x}_t$ 是在范围内 $[-H,H]$($\alpha=0$)或不在范围内($\alpha=1$)。状态 4 通过 c 和 α 来判断迭代是否成功。如果成功,则更新 $\boldsymbol{r}$, $x(j)$ 和索引 j。在同一个时钟周期内更新向量 $\boldsymbol{r}$ 的元素,因为它们已经被存储在寄存器中。另外,迭代计数器 p 加 1,并且 Flag 被设置为 1,这表明当前迭代是"成功"。系统也需要检查"成功"迭代计数器 p 是否达到最大数量 N_u,如果达到最大迭代次数,则算法停止。如果没有,算法判断是否更新 Flag,并根据 j 和 Flag 决定接下来将进入到哪一个状态。

表 7-6 显示了每个状态所需的时钟周期数。在每次迭代中,需要三个时钟周期。表 7-6 中显示的时钟周期上限可以看作是定点型盒型约束 DCD 算法并行架构在实现最坏情况时所需要的复杂度。当只有最后一位有 N_u 次成功迭代时才会出现最坏情况。这意味着前 (M_b-1) 位的计算不包含任何成功的迭代,因此需要 $(M_b-1)3K$ 个周期。计算最后一位($m=1$)的最坏情况是发生在 K 次迭代 $(j=1,\cdots,K)$ 中只发生一次成功迭代。这需要 $3K$ 个周期来进行比较,并且预更新残差向量 $\boldsymbol{r}$ 和元素 $x(j)$。N_u 次成功迭代总共需要 $3KN_u$ 个时钟周期。除了以上所需的周期外,整个过程中还需要 M_b 个时钟周期来更新步长。因此,定点型盒型约束 DCD 算法的时钟周期的上限为 $3KN_u+3K(M_b-1)+M_b$。

表 7-6 并行盒型约束 DCD 算法

状态	操作	周期
0	Initialization:$\boldsymbol{x}=0$, $\boldsymbol{r}=\boldsymbol{\theta}$, $m=M_b$, $p=0$, $j=1$, Flag$=0$	
1	if $m=0$, algorithm stops else, $m=m-1$, $d=2^m$, $r=2r$	1
2	$r_t=\boldsymbol{r}-\text{sign}(r(j))\cdot\boldsymbol{R}(:,\text{j})$ $c=R(j,j)/2-\|r(j)\|\times2^{\Delta m}$ $\boldsymbol{x}_t=x(j)+\text{sign}(r(j))\cdot d$ if $\|x_t\|\leqslant H$, then $\alpha=0$; else, $\alpha=1$	1

续表 7-6

状态	操作	周期
3	$c=R(j,j)/2-\lvert r(j)\rvert$ $x_t=x(j)+\text{sign}(r(j))\cdot d$ if $\lvert x_t\rvert\leq H$, then $\alpha=0$; else, $\alpha=1$	1
4	if $c<0$, and $\alpha=0$, $x(j)=x_t$ $\boldsymbol{r}=r_t$ $p=p+1$, Flag=1 if $p=N_u$, algorithm stops $j=(j)\bmod(K)+1$ if $j=1$ and Flag=1, then Flag=0, goto state 2 else if $j=1$ and Flag=0, then goto state 1 else, goto state 2	1
总共：	$\leq 3KN_u+3K(M_b-1)+M_b$	

***R*-in-Register**：矩阵 $\boldsymbol{R}$ 和向量 $\boldsymbol{r}$ 的元素存储在寄存器中（见图 7-8）。在状态 1，向量 $\boldsymbol{r}$ 左移一位和更新步长 d。在状态 2，读出向量 $\boldsymbol{r}$ 和 $\boldsymbol{R}(:,j)$ 中的元素来更新残差向量 $\boldsymbol{r}$。在状态 3，在一个时钟周期中比较 $r(j)$ 和 $R(j,j)$ 以及 $x(j)$ 的预更新。在状态 4，在一个时钟周期内更新向量 $\boldsymbol{r}$ 中所有元素和获取 $x(j)$。在这种情况下，由于矩阵 $\boldsymbol{R}$ 的所有元素都存储在寄存器中，所以芯片占用面积很大。为了进一步改进设计，我们考虑将 $\boldsymbol{R}$ 存储在一个 RAM 块中。

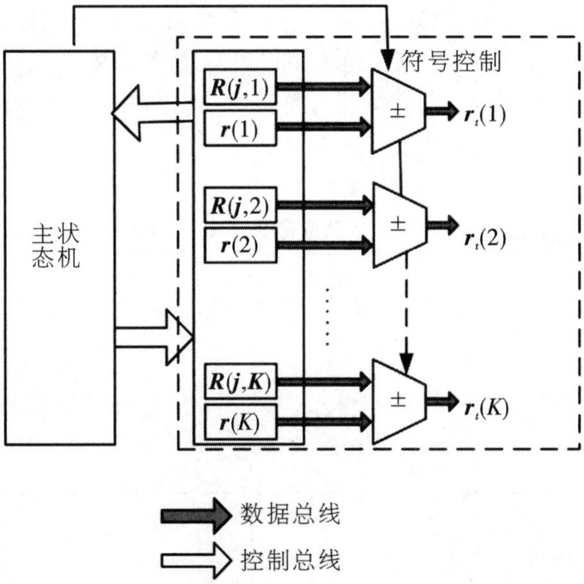

图 7-8　*R*-in-Register 形式的 *r* 更新器架构

R-in-RAM：矩阵 **R** 的行元素被存储在一组 RAM 中而不是寄存器。与 **R** 的元素存储在寄存器中的情况相比，**R**-in-RAM 方式可以明显减少硬件资源。这个设计如图 7–9 所示。

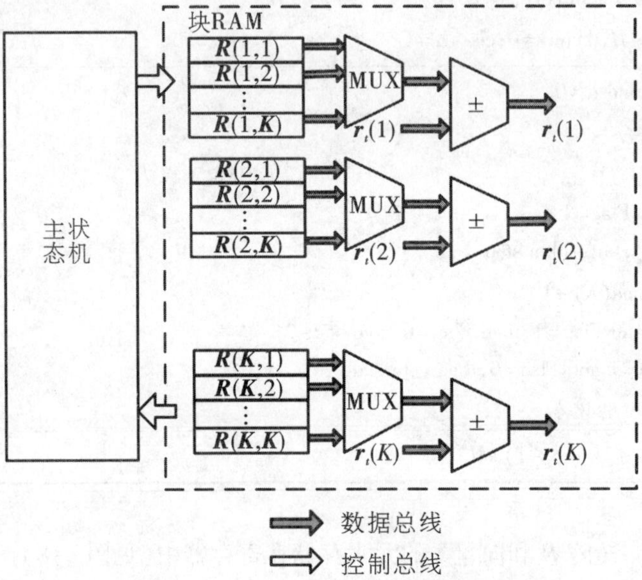

图 7–9　**R**-in-RAM 形式的 **r** 更新器架构

盒型约束 DCD 检测器的 FPGA 并行架构设计与串行架构设计相比，提高了吞吐量。然而，由于其占用的硬件资源较高，这两种并行架构设计更适用于较少用户数量的情况。表 7–7 列出了在 $K=16$ 和 $M_b=15$ 的情况下 **R**-in-Register 和 **R**-in-RAM 实现所需的 FPGA 资源。**R**-in-Register 执行操作所需 FPGA 资源高于 **R**-in-RAM 执行操作，因为它需要更多的寄存器。因此，**R**-in-Register 实现适用于小规模系统。与 **R**-in-Register 执行操作相比，**R**-in-RAM 执行操作减少了硬件芯片面积。然而，它与串行架构盒型约束 DCD 算法架构相比，它仍然需要较多的硬件资源。

表 7–7　$K=16$ 和 $M_b=15$ 的并行架构盒型约束 DCD 算法所需的 FPGA 资源

FPGA 资源	**R**-in-Register	**R**-in-RAM
逻辑片	7176	1465
D 触发器	5123	802
查询表	5646	2754
块 RAM	2	18

第 8 章　基于盒型约束的 DCD 复数 MIMO 检测

本章主要介绍基于盒型约束 DCD 算法 MIMO 系统检测方法和 FPGA 设计。以小规模 4×4 MIMO 系统为例，在相同检测性能前提下，该设计与 MMSE 算法相比在低 SNR 情况下需要的时钟周期更少。结合编码传输技术，DCD 算法可以在大规模 MIMO 系统中提供更优的检测结果。

8.1　概述

空间复用 MIMO 通信系统与单输入单输出（SISO）通信系统相比，可以增加信道容量。这需要在接收机上配置高效的检测技术。从实际情况出发，我们需要设计出 MIMO 检测器在硬件实现例如 FPGA 的合理方案。ML-MIMO 检测器可以提供最佳检测性能；然而，它的实时实现是非常复杂的。ML 检测器仅适用低阶调制的小规模 MIMO 系统（例如，信号用 QPSK 调制的 4 根发射天线和 4 根接收天线 MIMO 系统）的硬件实现。

球形译码器被作为最佳检测器硬件实现的一个最好选择。但是，随着系统规模和调制阶数的增加，该算法也变得更加复杂。另外，它在低信噪比下需要更多的运算量。文献[3,4]给出了球形解码器在 SNR 等于 20 dB 的吞吐量，在较低的 SNR 情况下，它的吞吐量显著降低。因此球形解码器通常实现于小规模 MIMO 系统（如 4×4 系统）的 FPGA 平台。此外，球形解码器在正式解码之前，需要 Cholesky 或 QR 的矩阵分解以及解相关检测。这些操作对于实时硬件操作是很困难的。

最佳检测器和它的次优近似算法（如球形解码器）只能提供硬判决。然而，软判决才是我们最感兴趣的，因为它们可以协助进一步有效的解码。软判决的 MIMO 检测通常是基于 MMSE 算法检测。最近，文献[8]-[10]论述了 MMSE MIMO 检测的不同 FPGA 设计方案。然而，随着天线数量的增加，MIMO 系统的硬件复杂度迅速增加，在 FPGA 平台上只能实现小规模 MMSE MIMO 检测器。而且，在大规模系统中，MMSE 检测器的性能明显劣于最佳检测的性能。

本章我们将学习一种软判决输出的盒型约束 MIMO 检测算法。我们已经知道 CDMA 系统中广泛使用了盒型约束多用户检测算法。盒型约束型算法与 MMSE 检测性能相比，可以提供更好的检测性能。并且，可以用 DCD 迭代来有效地实现 MMSE 算法。在第 7 章里，我们介绍了盒型约束 DCD 算法多用户检测器的 FPGA 设计。然而，该设计只能用于实数系统，例如，BPSK 调制信号和实数系统矩阵 $\boldsymbol{R}$。在 MIMO 系统中，我们用的更多是诸如 QPSK、16-QAM 等复数调制方案。在本章中，我们将介绍基于各种 QAM 调制的复数

DCD 盒型约束 MIMO 检测器的 FPGA 设计,并在硬件资源面积、吞吐量和检测性能等方面把它和 MMSE MIMO 检测器进行比较。

8.2 系统模型和盒型约束 MIMO 检测器

我们考虑 M_T 根发送天线和 M_R 根接收天线,具有瑞利平坦衰落信道的 $M_T \times M_R (M_T = M_R)$ MIMO 系统。接收信号是:

$$\boldsymbol{y} = \boldsymbol{G}\boldsymbol{x} + \boldsymbol{n} \tag{8.1}$$

其中 $\boldsymbol{G}$ 是 $M_R \times M_T$ 信道矩阵,矩阵里的元素是符合独立相同分布的(i. i. d.)零均值高斯随机数,$\boldsymbol{x}$ 是 $M_T \times 1$ QAM 星座 $\mathcal{A}$ 中的传输数据向量,而 $\boldsymbol{n}$ 是元素符号符合 i. i. d. 零均值高斯随机数的噪声向量。ML 检测器用整数约束来求解二次优化问题:

$$\begin{aligned}\hat{\boldsymbol{x}}_{\mathrm{ML}} &= \arg \min_{\boldsymbol{x} \in \mathcal{A}^{M_T}} \{ \| \boldsymbol{y} - \boldsymbol{G}\boldsymbol{x} \|^2 \} \\ &= \arg \min_{\boldsymbol{x} \in \mathcal{A}^{M_T}} \{ \boldsymbol{x}^{\mathrm{H}} \boldsymbol{R} \boldsymbol{x} - 2\,\boldsymbol{\theta}^{\mathrm{H}} \boldsymbol{x} \}\end{aligned} \tag{8.2}$$

其中 $\boldsymbol{R} = \boldsymbol{G}^{\mathrm{H}}\boldsymbol{G}$ 和 $\boldsymbol{\theta} = \boldsymbol{G}^{\mathrm{H}}\boldsymbol{y}$ 。盒型约束检测器放松约束条件 $\boldsymbol{x} \in \mathcal{A}^{M_T}$、$\Re\{\boldsymbol{x}\} \in [-H, H]^{M_T}$ 和 $\Im\{\boldsymbol{x}\} \in [-H, H]^{M_T}$,其中 H 是调制星座图 $\mathcal{A}$ 中元素的实部和虚部最大值。例如,对于 16-QAM 调制, $H = 3$。因此,得到盒型约束 MIMO 检测器的解

$$\hat{\boldsymbol{x}}_{\mathrm{box}} = \arg \max_{\Re\{\boldsymbol{x}\} \& \Im\{\boldsymbol{x}\} \in [-H,H]^{M_T}} \{ \boldsymbol{x}^{\mathrm{H}} \boldsymbol{R} \boldsymbol{x} - 2\,\boldsymbol{\theta}^{\mathrm{H}} \boldsymbol{x} \} \tag{8.3}$$

该解决方案提供软输出 $\hat{\boldsymbol{x}}_{\mathrm{box}}$,然后把软输出结果映射到 $\mathcal{A}^{M_T}$ 中。求解(8. 3)等价于求解盒型约束正则方程组

$$\boldsymbol{R}\boldsymbol{x} = \boldsymbol{\theta} \tag{8.4}$$

中的盒型约束解 $\boldsymbol{x}$ 。

8.3 基于 DCD 的盒型约束 MIMO 检测器的 FPGA 实现

DCD 算法采用 2 的不同指数幂作为步长的坐标下降迭代法求解正则方程(8. 4)。迭代通常有两种类型:成功和不成功。在成功迭代中,更新解向量 $\boldsymbol{x}$ 的一个元素和残差矢量 $\boldsymbol{r} = \boldsymbol{\theta} - \boldsymbol{R}\boldsymbol{x}$ 的所有元素; 这些成功迭代在算法中所需的运算量最大。在不成功迭代中,没有更新。成功迭代次数(更新) N_u 是预先定义的。另一个参数 M_b 也是预先定义的,它表示 $\boldsymbol{x}$ 里元素的位数,因此可以控制结果的最终精度。它还决定了步长 d 被除以 2 的次数。

表 8 - 1 是盒型约束 DCD 算法针对求解复数调制信号的 FPGA 优化实施。图 8 - 1 是 DCD 处理器的架构框图。

表8-1　复数盒型约束DCD算法

状态	操作	周期
0	Initialization: $x=\bar{x}, \boldsymbol{r}=\boldsymbol{\theta}-\boldsymbol{Rx}, m=M_b, p=0$, $\Delta m=0, s=1, j=1$, Flag $=0$	
1	if $m=0$, algorithm stops else, $m=m-1, d=2^{m}, \Delta m=\Delta m+1$	1
2	if $s=1$, then $r_t=\Re(r(j))$; else, $r_t=\Im(r(j))$ $c=R(j,j)/2-\lvert r(j)\rvert\times 2^{\Delta m}$ $x_t=x(j)+\mathrm{sign}(r_t)\cdot s\cdot d$ if $\lvert\Re(x_t)\rvert\leqslant H$ and $\lvert\Im(x_t)\rvert\leqslant H$, then $\alpha=0$; else $\alpha=1$	1
3	if $c<0$, and $\alpha=0$, goto state 4 else, goto state 5	1
4	$x(j)=x_t$ $r=r\times 2^{\Delta m}-\mathrm{sign}(r_t)s\boldsymbol{R}^{(j)}$ $\Delta m=0, p=p+1$, Flag $=1$ if $p=N_u$, algorithm stops	M_T
5	if $s=1$, then $s=i$, goto state 2 else, $s=1, j=(j)\mathrm{mod}(M_T)+1$ if $j=1$ and Flag $=1$, then Flag $=0$ goto state 2 else if $j=1$ and Flag $=0$, then goto state 1 else, goto state 2	1
总共:	$\leqslant 7M_T N_u+6M_T(M)_b-1)+M_b$周期数	

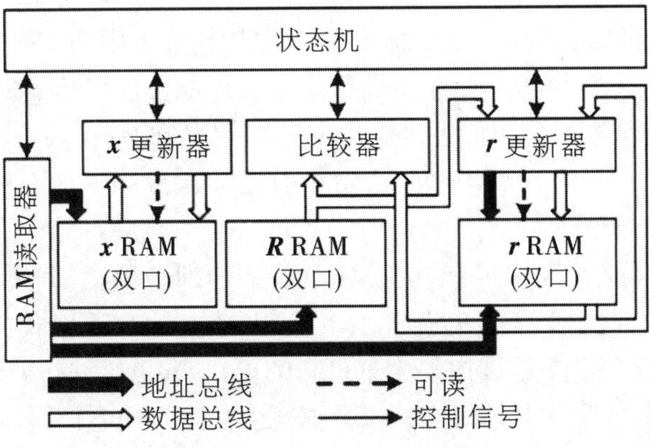

图8-1　DCD处理器的框图

DCD 状态机控制整个算法。如表 8-1 所示，在状态 0 中，几个控制信号进行初始化。残差向量是 $\boldsymbol{r}=\boldsymbol{\theta}-\boldsymbol{R}\bar{\boldsymbol{x}}$，其中 $\bar{\boldsymbol{x}}$ 是 $\boldsymbol{x}$ 的初始值。在这一章中，$\bar{\boldsymbol{x}}$ 被设置为零，则 $\boldsymbol{r}$ 等于 $\boldsymbol{\theta}$。比特计数器 m 被设置为 M_b，成功迭代计数器 p 被设置为 0，预设的缩放计数器 Δm 被设置为 0，当计算元素 $r(j)$ 的实部时 s 设置为 1，计算元素 $r(j)$ 的虚部时 s 设置为 i，元素索引 j 设置为 1，成功迭代指示符“Flag”设置为 0。

在状态 1 中，更新位计数器 m，步长 $d=2^m$ 和预设的缩放计数器 Δm。$m=0$ 意味着已经确定了结果中元素的最低有效位，DCD 处理器停止运算；否则，算法进入状态 2。

在状态 2 中，RAM 读取器确定 $\boldsymbol{\theta}$ RAM 中 $r(j)$ 地址，$\boldsymbol{R}$ RAM 中 $R(j,j)$ 地址，$\boldsymbol{x}$ RAM 中 $x(j)$ 地址。然后，根据 s 的数值，比较器选择 $r(j)$ 的实部（如果 $s=1$）或虚部（如果 $s=i$）。如果向量 $\boldsymbol{r}$ 的第 m 位预先没有被缩放，比较器就会缩放 r_t。然后进行比较并将结果 c 的符号位发送给 DCD 状态机。同时，$\boldsymbol{x}$ 更新器从 $\boldsymbol{x}$ RAM 中读取 $x(j)$，根据 s 的数值，预先更新 $x(j)$ 的实数或虚数分量，把更新后的元素作为 x_t 保存在寄存器中，并检查其是否在范围 $[-H,H]$（$\alpha=0$）内，不在范围内则 $\alpha=1$，H 的大小取决于调制，例如，M-PSK 信号 $H=1$。

在状态 3 中，DCD 状态机检查比较结果 c 和 α 的值来决定将进行到哪一步骤。如果 $c<0$ 且 $\alpha=0$（迭代成功），则算法进入状态 4 更新 $x(j)$ 和 $\boldsymbol{r}$；否则，不更新进入状态 5（迭代不成功）。

在状态 4 中，$\boldsymbol{r}$ 更新器更新 $\boldsymbol{r}$。RAM 读取器在 $\boldsymbol{R}$ RAM 和 $\boldsymbol{\theta}$ RAM 中分别生成列向量 $\boldsymbol{R}(:,j)$ 和向量 $\boldsymbol{r}$ 中元素的地址。$\boldsymbol{r}$ 更新器从 $\boldsymbol{\theta}$ RAM 中读取 $\boldsymbol{r}$ 的元素，进行更新，并将结果写回 $\boldsymbol{\theta}$ RAM。$\boldsymbol{r}$ 更新器有两个加法器可以同时更新 $\boldsymbol{r}$ 的实部和虚部。$\boldsymbol{x}$ 更新器将 $\boldsymbol{x}_t$ 写入 $\boldsymbol{x}$ RAM 取代 $x(j)$。然后，DCD 状态机将预设的缩放计数器 Δm 设置为 0，把 Flag 设置为 1 来表示迭代成功。成功迭代的计数器 p 也被更新；如果 p 等于预定限制值 N_u，则 DCD 处理器停止；否则，进入状态 5。

在状态 5 中，DCD 状态机首先检查 s 的值以确定接下来应该分析 $\boldsymbol{x}$ 的哪个分量。然后，更新索引 j 和信号 s。最后，更新 Flag，并根据 j 和 $Flag$ 进入下一个状态。

表 8-1 显示了每个状态所需的时钟周期数。在状态 4 中采用流水线方式，需要 M_T 个时钟周期来更新 $x(j)$ 和 $\boldsymbol{r}$ 中的所有元素。其他每个状态都需要一个周期。求解方程组的周期数取决于系统规模、系统矩阵条件件数以及算法参数 N_u 和 M_b。对于给定的 M_{T}、N_u 和 M_b，它所需要的时钟周期数是对应最坏情况周期数上限的随机数，它的周期数上限是 $7M_{\mathrm{T}}N_u+6M_{\mathrm{T}}(M_b-1)+M_b$，对于更高数值 N_u，周期数近似等于 $7M_{\mathrm{T}}N_u$。周期上限对应的情况，即在确定结果的每一位多次迭代中只有一次成功迭代（一次更新）。在通常情况下，确定结果的每一位的一组迭代中通常会有很多次成功迭代。所以，时钟周期平均数会比上限值要小。

复数方程组的每个元素用两个 16 位 Q15 数字分别表示实部和虚部，实部和虚部数值是在 $[-1,1)$ 范围内。为了避免溢出，采用 32 位定点型 Q15 格式存储 $\boldsymbol{r}$ 和 $\boldsymbol{x}$ 的实部和虚部，并限制在 $[-2^{16},2^{16})$ 范围内。为了更高的更新率，可以并行处理 $\boldsymbol{r}$ 的实部和虚部。$\boldsymbol{R}$ RAM 是 32 位数据宽度，$\boldsymbol{r}$ RAM 是 64 位数据宽度。这可使它们同时支持实部和虚部的读写操作。由于每次迭代仅需 $x(j)$ 的实部或虚部，$\boldsymbol{x}$ RAM 是 32 位数据宽度。

表8-2给出了盒型约束DCD的MIMO复数检测器所需的FPGA资源。可以看出,盒型约束DCD算法的硬件面积使用量非常小,求解 $M_T=4$ 到 $M_T=16$ 的方程组,分别需要637到667个逻辑片。在所有情况下,设计所占用的芯片面积都小于5%,并且不使用任何内置的乘法器。与文献[8]-[10]中所提出的分别需要8513、7679和9474个逻辑片的 4×4MMSE MIMO检测器设计相比,盒型约束DCD所需要的硬件面积显著减少。另外,文献[8]、[9]中提出的MMSE检测器分别使用了占用较大硬件面积的64个和58个乘法器,而DCD算法中的乘法器数量为零。

表8-2 盒型约束DCD的MIMO复数检测器所需的FPGA资源

资源	$M_T=4$	$M_T=8$	$M_T=16$
逻辑片	637(4.7%)	658(4.8%)	667(4.9%)
D触发器	305(1.1%)	318(1.2%)	329(1.2%)
查询表	1033(3.8%)	1062(3.9%)	1084(4.0%)
块RAM	5(3.7%)	5(3.7%)	5(3.7%)

8.4 数值结果

在AWGN信道中,我们用数值结果来估计所提出设计的吞吐量。具体地,是根据更新次数和时钟周期数目来显示16-QAM调制的 4×4、8×8 和 16×16 MIMO系统的收敛速度。

通过求解(8.3),DCD算法($M_b=15$)获得解 $\hat{\boldsymbol{x}}_{\text{box}}$。通过下式来计算估计值 $\hat{\boldsymbol{x}}_{\text{box}}$ 和发送向量 $\boldsymbol{x}$ 之间的误差:

$$\xi=\frac{\|\hat{\boldsymbol{x}}_{\text{box}}-\boldsymbol{x}\|^2}{\|\boldsymbol{x}\|^2}. \tag{8.5}$$

平均误差通过 $T=1000$ 次模拟试验,以分贝形式给出:

$$\bar{\xi}=10\lg\left\{\frac{1}{T}\sum_{t=1}^{T}\frac{\sum_{j=1}^{M_T}\left|\hat{x}_{\text{box}}(j)-x(j)\right|^2}{\sum_{j=1}^{M_T}\left|x(j)\right|^2}\right\} \tag{8.6}$$

图8-2是用MATLAB获得的不同更新数量 N_u 下的误差结果图。

为了比较,我们还显示了用DCD算法实现的MMSE MIMO检测器的结果。MMSE检测器算法和盒型约束检测器不同,表8-1中它去除了状态2和状态3的阈值 H 比较,用 $\boldsymbol{R}+\frac{1}{\text{SNR}}\boldsymbol{I}$ 代替矩阵 $\boldsymbol{R}$,其中 $\boldsymbol{I}$ 是 $M_T\times M_T$ 单位矩阵。盒型约束解决方案比MMSE解决方案误差更低,而且两者之间的性能差距随着系统规模 M_T 的增加而增加。

图 8-3 显示了随 FPGA 周期数变化的误差趋势。固定一个误差数值例如-25 dB 来比较图 8-3 和图 8-2,我们可以得出结论,对于 4 × 4、8 × 8 和 16 × 16 MIMO 系统来说,平均一次更新所需要的时间分别近似 $2.5M_T$、$2M_T$ 和 $1.7M_T$ 个时钟周期。这个数值明显低于上面讨论的最坏情况下的时钟周期数。因此,对于一个 4 × 4 MIMO 系统,所需的时钟总数近似为 $2.5M_TN_u$,对于一个 8 × 8 MIMO 系统,所需的时钟总数近似为 $2M_TN_u$,对于一个 16 × 16 MIMO 系统,所需的时钟总数近似为 $1.7M_TN_u$.

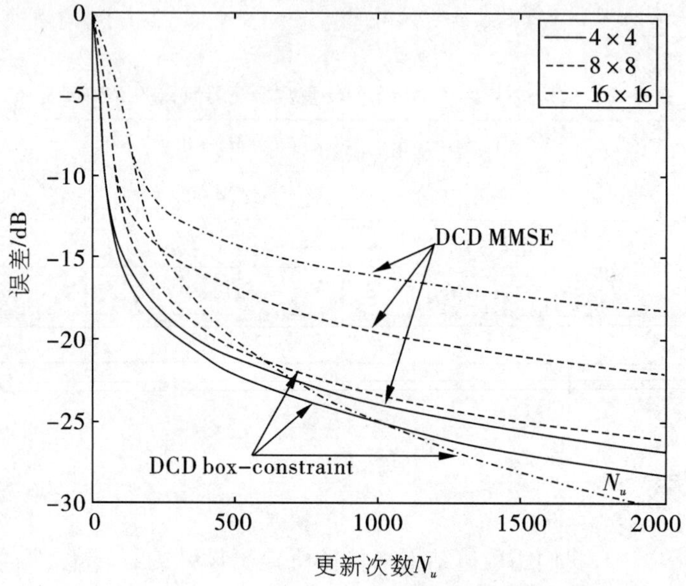

图 8-2　在 16-QAM MIMO 系统中误差与更新次数 N_u

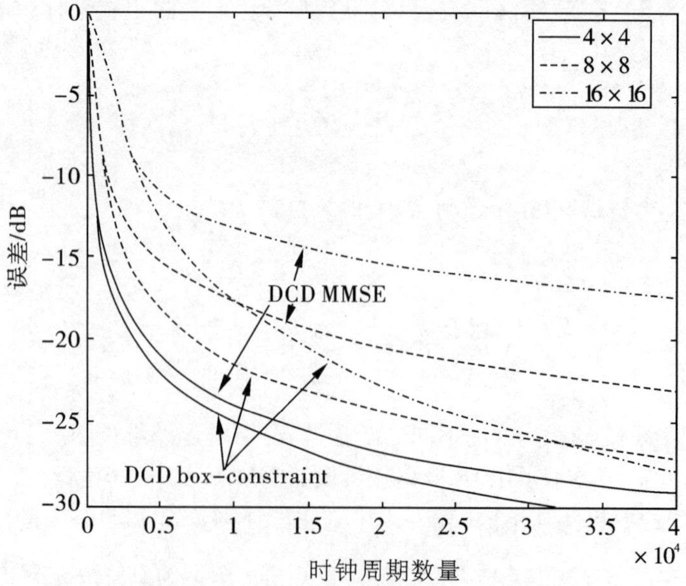

图 8-3　在 16-QAM MIMO 系统中误差与时钟周期数量

图 8-4、图 8-5 和图 8-6 比较了在 4 × 4、8 × 8 和 16 × 16 MIMO 系统中基于 DCD 的 MMSE 检测器和典型 MMSE 检测器浮点型 BER 性能。这些结果表明，盒型约束检测器性能显著优于 MMSE 检测器，特别是对于大规模 MIMO 系统。对于固定的 N_u，DCD 迭代的检测器呈现了 BER 下限值，这个下限值随着 N_u 增加而减小。

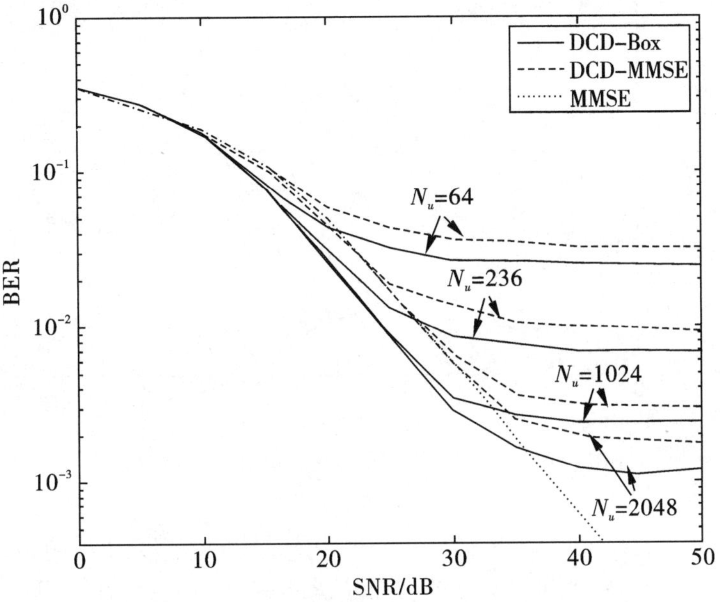

图 8-4　4 × 4 MIMO 系统检测器的 BER 性能

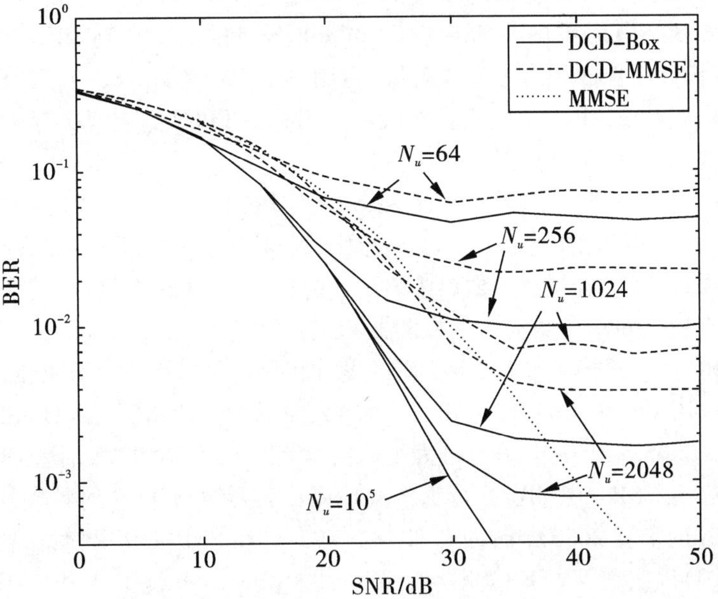

图 8-5　8 × 8 MIMO 系统检测器的 BER 性能

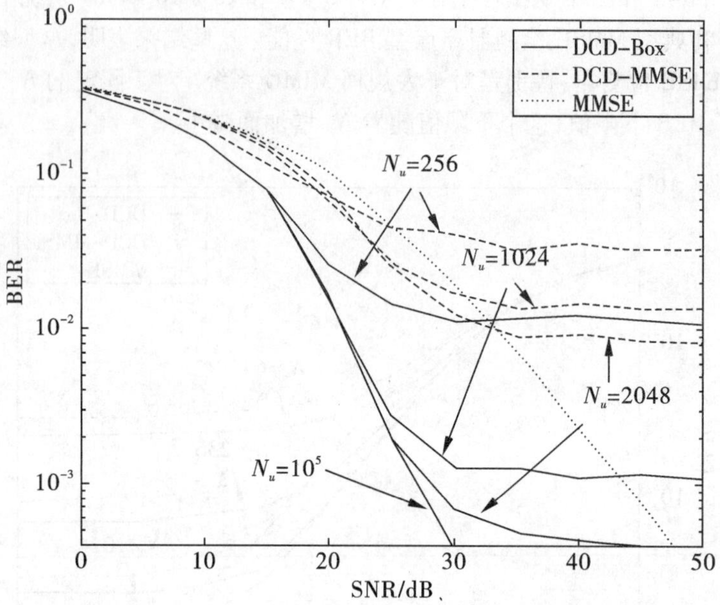

图 8-6　16 × 16 MIMO 系统检测器的 BER 性能

图 8-7 显示了在 4 × 4、8 × 8 和 16 × 16 MIMO 系统中盒型约束检测器的 BER 性能随更新数量的增加而改善。对于 4 × 4 的情况，从图 8-4 可以看出，在 SNR = 10 dB，20 dB 和 30 dB 时，MMSE 检测器的 BER 分别为 0.18、0.05 和 0.006。图 8-7 表明，盒型约束检测器分别用 N_u = 27、47 和 410 取得一样的性能分别需要 270、470 和 3800 个周期。我们可以得出这样的结论：从吞吐量的角度来看(每个时钟周期所处理数据量)，盒型约束 DCD 设计不如[8]-[10]中的 MMSE 设计，MMSE 设计需要 270 到 388 个时钟周期。然而，由于 DCD 设计硬件面积占用率非常低(占用少于 5% 的 XC2VP30(FFT896 封装，速度等级 7)的 Xilinx Virtex-II Pro 开发系统)，可以用整个 FPGA 芯片实现大约 20 个盒型约束 DCD 检测器。通过 20 个盒型约束检测器并行工作，可以在 SNR = 10 dB，20 dB 和 30 dB 把检测一个 MIMO 符号的平均周期数减少到 14、24 和 190 个时钟周期，明显低于 MMSE 检测器。如果使用更先进的 FPGA 芯片，例如 Virtex-5 XCE5VLX330，可以把并行处理的 DCD-MIMO 检测器数量增加到 200。注意，在 OFDM MIMO 系统中，实现多个检测器并行工作(例如每个子载波一个检测器)是有益的。

随着 MIMO 系统规模的增加，所提出的设计将显得更有效。特别地，对于 8×8 MIMO 系统，在 SNR = 10 dB、20 dB 和 30 dB，N_u = 40、68 和 297 的情况下，MMSE 检测器取得 BER 性能(BER = 0.22、0.07 和 0.01)大约需要 650、1100 和 4800 个时钟周期。对于 16×16 MIMO 系统，在 SNR = 10 dB 时，盒型约束检测器需要约 1800 个时钟周期(N_u = 68)，在 SNR = 20 dB 时需要 2600 个时钟周期(N_u = 95)，在 SNR = 30 dB 时需要 4600 个时钟周期(N_u = 170)以实现 MMSE 检测器的 BER 性能(BER = 0.25、0.1 和 0.02)。表 8-3 显示了所有时钟周期数目。可以看出，所提出的设计特别适用于大规模 MIMO 系统。

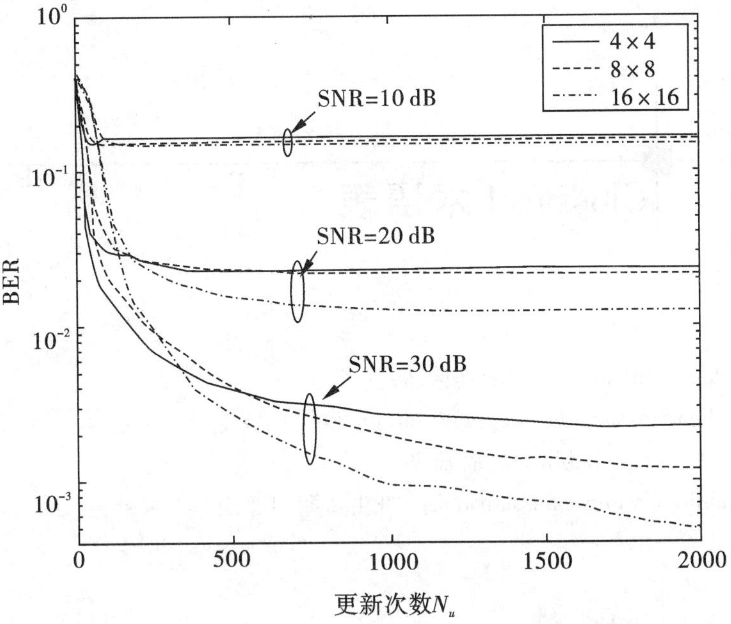

图 8-7　16-QAM MIMO 系统中不同 N_u 下的 BER 性能

表 8-3　实现 MMSE 检测器相同 BER 性能的盒型约束 DCD MIMO 检测器所需的周期数

SNR	4 × 4	8 × 8	16 × 16
10 dB	270	650	1800
20 dB	470	1100	2600
30 dB	3800	4800	4600

Glossary(术语表)

A

ADC:analog to digital converter 模数转换器

ASIC:application specific integrated circuit 专用集成电路

APP:a posteriori probability 后验概率

AWGN:additive white Gaussian noise 加性高斯白噪声

B

BS:base station 基站

BPSK:binary phase shift keying 二进制相移键控

BER:bit error rate 误码率

BB:branch and bound 分支定界算法

BAMI:block analysis matrix inverse 分块分析矩阵求逆

C

CDMA:code division multiple access 码分多址

CORDIC:coordinated rotation digital computer 坐标旋转数字计算机

CVP:closest vector problem 最近向量问题

CG:conjugate gradient 共轭梯度

D

DSSS:direct sequence spread spectrum 直接序列扩频

DS:direct sequence 直接序列

DAC:digital to analog converter 数模转换器

DSP:digital signal processing 数字信号处理

DCD:dichotmous coordinate descent 二分坐标下降

DF:decision feedback 判决反馈

F

FHSS:frequency-hopping spread spectrum 跳频扩频

FFHSS:fast frequency-hopping spread spectrum 快速跳频扩频

FBMC:filter bank multicarrier 滤波器组多载波
FSK:frequency-shift keying 频移键控
FDMA:frequency division multiple access 频分多址
FBMC:filter bank multi-carrier 滤波器组多载波
FPGA:field programmable gate array 现场可编程门阵列
FEC:forward error correction 前向纠错

G

GPS:global position system 全球定位系统

H

Hz:Hertz 赫兹

I

ISI:inter-symbol interference 码间干扰
LNS:logarithmic number system 对数数字系统
I. I. D:independent identically distributed 独立同分布
IC:interference cancellation 干扰消除

L

LFHSS:low frequency-hopping spread spectrum 慢速跳频扩频
LTE:long term evolution 长期演进
LOS:line of sight 视线
LS:least square 最小二乘法
LAPACK:linear algebra package 线性代数程序包

M

MAI:multiple access interference 多址干扰
MIMO:multiple input multiple output 多输入多输出
MUD:muitiuser detection 多用户检测
ML:maximum likelihood 最大似然
MF:matched filter 匹配滤波器
MMSE:minimum mean square error 最小均方误差
MRC:maximum ratio combining 最大比合并
MPSK:M - ary phase-shift keying 多进制相移键控

N

NIOS:netware input/output subsystem 网件输入/输出子系统

O

OSIC：ordered successive interference cancellation 排序连续干扰消除

OFDM：orthogonal frequency division multiplexing 正交频分复用

P

PRN：pseudo random number 伪随机数

PA：power amplifier 能量放大器

PBX：private branch exchange 专用分组交换机

PDF：probability density function 概率密度函数

PDA：probabilistic data association 概率数据关联

P-DF：DF receiver with parallel interference cancellation 并行干扰消除判决反馈接收器

Q

QoS：quality of service 服务质量

QRD：QR decomposition QR 分解

QPSK：quadrature phase shift keying 正交相移编码

QAM：quadrature amplitude modulation 正交振幅调制

QP：quadratic programming 二次方程

R

RF：radio frequency 射频

RMS：root mean square 均方根

RAM：random access memory　随机存取存储器

RLS：recursive least squares　递归最小二乘

S

SS：spread spectrum 扩频

SISO：single-input single-output 单输入单输出

SIMO：single-input multiple-output 单输入多输出

SGR：squared Givens rotations 吉文斯旋转平方

SOR：successive over-relaxation 逐次超松驰

SD：sphere decoding 球形译码

SNR：signal noise ratio 信噪比

SER：symbol error rate 误码率

SAC：shift-accumulation 移位累加

S-DF：DF detector with successive interference cancellation 串行干扰消除判决反馈接收器

SDR:semidefinite relaxation 半正定松弛
SM:spatial multiplexing 空间复用
STC:space-time coding 空时编码
STTC:space time trellis coding 空时网格编码
STBC:space time block coding 空时分组编码

T

THSS:time-hopping spread spectrum 跳时扩频
TDMA:time division multiple access 时分多址

U

UMTS:universal mobile telecommunications system 通用移动通信系统

V

VLSI:very large scale integration circuit 超大规模集成电路
VHDL:hardware description language 硬件描述语言

W

WCDMA:wideband code division multiple access 宽带码分多址
W-LAN:wireless local area network 无线局域网

Z

ZF:zero forcing 迫零

Notation（符号表）

符号	含义	
$R(i,j)$	(i,j) entry of R	R 的第 (i,j) 个元素
$R(:,j)$	jth coloum of R	R 的第 j 列
R^{T}	transpose of R	R 的转置
R^{H}	Hermitian transpose of R	R 的共轭转置
$\det(A)$	determinant of A	A 的行列式
$\mathrm{rank}(A)$	rank of A	A 的秩
$\Re\{\cdot\}$	real component	实数部分
$\Im\{\cdot\}$	imaginary component	虚数部分
$\boldsymbol{I}$	identity matrix	单位矩阵
e_i	ith column of the identity matrix	单位矩阵第 i 列
$E\{\cdot\}$	expection	期望